The Geography of Soils

The Geography of Soils

Formation, Distribution and Management

2nd Edition

DONALD STEILA
and
THOMAS E. POND

Rowman & Littlefield Publishers, Inc.

ROWMAN & LITTLEFIELD PUBLISHERS, INC.

Published in the United States of America in 1989
by Rowman & Littlefield, Publishers, Inc.
8705 Bollman Place, Savage, Maryland 20763

Library of Congress Cataloging-in-Publication Data

Steila, Donald, 1939-
    The geography of soils : formation, distribution, and management /
Donald Steila, Thomas E. Pond.—2nd ed.
        p. 261
    Bibliography: p.
    Includes index.
    ISBN 0-8476-7592-0
    1. Soil geography. 2. Soil science. I. Pond, Thomas E.
II. Title.   1989   88-11446
S591.S834   CIP
631.4—dc19

Printed in the United States of America

To my parents, Mary and John Steila, Sr.,
and
to my children, Stephanie and John F. Steila

Contents

Preface   *xix*

Introduction   *1*

1

**The Origin and Significance
of the Soil's Inorganic Constituents**   *3*

   Minerals and Parent Material
   Products of Weathering

2

The Organic Fraction of the Soil   *25*

   Soil Organisms
   Soil Organic Matter
   Organic Material and Soil Color

3

Soil Porosity, Moisture, and Atmosphere   *41*

   Pore Space
   Soil Moisture
   Soil Atmosphere

4

Effect of Site and Time on Soil Characteristics   *53*

Topographic Position
The Soil Profile: A Factor of Time

5

Soil Classification   *67*

U. S. Comprehensive Soil Classification System

6

Entisols, Vertisols, and Inceptisols   *83*

Entisols
Vertisols
Inceptisols
Land Utilization and Management Problems

7

Aridisols   *99*

Climate and Native Vegetation
Pedogenesis
Land Utilization and Management Problems

8

Mollisols   *115*

Climate and Native Vegetation
Pedogenesis
Land Utilization and Management Problems

9

Spodosols   *129*

Climate and Native Vegetation
Pedogenesis
Land Utilization and Management Problems

10

Alfisols and Ultisols   *143*

> *Pedogenesis*
> *Alfisols*
> *Ultisols*
> *Land Utilization and Management Problems*

11

Oxisols   *159*

> *Climate and Native Vegetation*
> *Pedogenesis*
> *Land Utilization and Management Problems*

12

Histosols   *173*

> *Land Utilization and Management Problems*

Appendices   *179*

> *I. Structure of Clay Minerals*
> *II. Descriptive Soil Profile Symbols*
> *III. Soil Color*

Glossary   *191*

Bibliography   *229*

Index   *235*

List of Tables

Table 1.1 Selected Original and Secondary Minerals

Table 1.2 Soil Separates

Table 1.3 Weathering Products of Selected Common Minerals

Table 4.1 Physical and Physio-chemical Properties of Horizon Soil Samples of the Recognized Soil Series in the Machkund Basin

Table 4.2 Master Horizons

Table 5.1 Soil Categories (Marbut, 1938)

Table 5.2 Soil Classification in the Higher Categories (revised, 1949)

Table 5.3 Soil Order's Name Derivation, Areal Significance, and Marbut Equivalents

Table 5.4 Formative Elements in Names of Suborders

Table 5.5 Formative Elements in Names of Great Groups

Table 5.6 Profile Description of an Umbric Paleaquult

Table 6.1 Profile Description of a Psamment

Table 7.1 Significant Features of Soil Classification in the Aridisol Order

Table 7.2 Precipitation Data for the Period 1951–1960 and Precipitation Normal Values for Las Vegas, Nevada

Table 7.3 Profile Description of Soil with Petrocalcic Horizon

Table 8.1 Profile Description of a Mollisol

Table 9.1 Profile Description of a Typic Sideraquod

Table 9.2 Yield Increases of Selected Crops After an Application of Two Metric Tons/Hectare of Limestone per Six-Year Rotation

Table 10.1 Profile Description of an Albaqualf

Table 11.1 Profile Description of an Oxisol in Mysore State, India

Table 12.1 Profile Description of a Borohemist (Chase County, Michigan)

List of Figures

Figure 1.1 Percentages of the common elements in the earth's crust by weight: O = oxygen, Si = silicon, Al = aluminum, Fe = iron, Ca = calcium, Na = sodium, K = potassium, and Mg = magnesium.

Figure 1.2 Examples of the difference between sedentary and transported parent material.

Figure 1.3 Relationship of the size of fragments to total surface area.

Figure 1.4 Resistance of minerals to weathering.

Figure 1.5 Accumulation of mineral constituents as weathering occurs. (The silicon is the mineral portion contained in silicate minerals; it is not quartz.)

Figure 1.6 Relationship between particle size and types of minerals present.

Figure 1.7 Conditions in which the various silicate clays and oxides of iron and aluminum may form.

Figure 1.8 Chart showing the percentage of clay (below 0.002 mm), silt (0.002 mm to 0.05 mm), and sand (0.05 mm to 2 mm) in the basic soil textural classes.

Figure 1.9 The common manner in which soil separates are arranged into structural units and their size class.

Figure 1.10 Diagram of a clay colloid (micelle) with its sheetlike morphology, numerous negative charges, and swarm of absorbed cations.

Figure 1.11 The pH scale to measure the concentration of hydrogen cations (H^+) in the soil.

Figure 2.1 Some important soil organisms.

Figure 2.2 Selected bacterial groups and their functions.

Figure 2.3 Relationship of plant communities to the depth of soil moisture penetration in the Great Plains of North America.

Figure 2.4 Organic matter accumulation bears an inverse relationship to mean annual temperature. Organic matter decomposition rates, however, increase with increased temperature.

Figure 3.1 Pore space and bulk density vary inversely. As inorganic soil components decrease in size, the total pore space normally increases and bulk density decreases.

Figure 3.2 Examples of the calculation of bulk density and particle density.

Figure 3.3 Forms of soil moisture. The amount of water available to plants is equal to the field capacity minus the wilting point.

Figure 3.4 General relationship between soil moisture availability and soil texture.

Figure 3.5 Water budgets of Brevard, North Carolina, and San Francisco, California.

Figure 3.6 Moisture regions in the United States (after C. W. Thornthwaite).

Figure 3.7 Seasonal variation in effective moisture (after C. W. Thornthwaite).

Figure 4.1 Relationship of elevation with precipitation and mean annual temperature in the Catalina Mountains of southern Arizona.

Figure 4.2 Relationship between elevation, precipitation, and vegetation.

Figure 4.3 A soil catena in a hypothetical landscape.

Figure 4.4 A sequence of soil profiles from youthful through mature.

Figure 4.5 An example of a possible horizon sequence.

Figure 4.6 The pedon in its three-dimensional form.

Figure 4.7 Soil Series as they relate to landscape position.

Figure 5.1 Great Soil Groups of the United States according to the Marbut classification scheme.

Figure 5.2 Relationship between soil orders and intensity of weathering. (The percentages provide the approximate areal extent of each Soil Order.)

Figure 5.3 World distribution of soils (courtesy of the Soil Conservation Service.)

Figure 6.1 World distribution of Entisols.

Figure 6.2 Entisols of the United States. (From Philip J. Gersmehl, "Soil Taxonomy and Mapping," *Annals of the Association of American Geographers,* 67, September 1977, p. 423. By permission of the Association of American Geographers.)

Figure 6.3 Suborders and Great Groups of the Soil Order Entisol.

Figure 6.4 World distribution of Vertisols.

Figure 6.5 Vertisols of the United States. (From Philip J. Gersmehl, "Soil Taxonomy and Mapping," *Annals of the Association of American Geographers,* 67, September 1977, p. 424. By permission of the Association of American Geographers.)

Figure 6.6 Vertisol Development. (From H.D. Foth, *Fundamentals of Soil Science,* 7th ed., New York: John Wiley and Sons, Publishers, 1984, p. 281.)

Figure 6.7 Suborders and Great Groups of the Soil Order Vertisol.

Figure 6.8 World distribution of Inceptisols.

Figure 6.9 Inceptisols of the United States. (From Philip J. Gersmehl, "Soil Taxonomy and Mapping," *Annals of the Association of American Geographers,* 67, September 1977, p. 423. By permission of the Association of American Geographers.)

Figure 6.10 Suborders and Great Groups of the Soil Order Inceptisol.

Figure 7.1 World distribution of Aridisols.

Figure 7.2 Aridisols of the United States. (From Philip J. Gersmehl, "Soil Taxonomy and Mapping," *Annals of the Association of American Geographers,* 67, September 1977, p. 424. By permission of the Association of American Geographers.)

Figure 7.3 Water budget of Las Vegas, Nevada (based on normal data).

Figure 7.4 Relationship between temperature and water vapor—holding capacity of the atmosphere.

Figure 7.5 A schematic model of Caliche development.

Figure 7.6 Suborders and Great Groups of the Soil Order Aridisol.

Figure 7.7 Soil Associations—Diagrammatic: Beryl-Enterprise Area, Utah. (Unpublished manuscript of classroom exercises, courtesy of Neil E. Salisbury.)

Figure 7.8 World distribution of stock raising and nomadic herding activities.

Figure 8.1 World distribution of Mollisols.

Figure 8.2 Mollisols of the United States. (From Philip J. Gersmehl, "Soil Taxonomy and Mapping," *Annals of the Association of American Geographers,* 67, September 1977, p. 425. By permission of the Association of American Geographers.)

Figure 8.3 Fluctuations in the humid/subhumid boundary in the Great Plains of the United States.

Figure 8.4 Water budget of Limon, Colorado (based on normal data).

Figure 8.5 Suborders and Great Groups of the Soil Order Mollisol.

Figure 8.6 Soil Associations—Diagrammatic of Polk County, Iowa (Unpublished manuscript of classroom exercises, courtesy of Neil E. Salisbury).

Figure 9.1 World distribution of Spodosols.

Figure 9.2 Spodosols of the United States. (From Philip J. Gersmehl, "Soil Taxonomy and Mapping," *Annals of the Association of American Geographers,* 67, September 1977, p. 425. By permission of the Association of American Geographers.)

Figure 9.3 Water budget of Mistassini Post, Quebec (based on normal data).

Figure 9.4 Suborders and Great Groups of the Soil Order Spodosol.

Figure 9.5 Northern Appalachian composite soil associations—Clarion and Potter Counties, Pennsylvania (Unpublished manuscript of classroom exercises, courtesy of Neil E. Salisbury).

Figure 10.1 World distribution of Alfisols.

Figure 10.2 Alfisols of the United States. (From Philip J. Gersmehl, "Soil Taxonomy and Mapping," *Annals of the Association of American*

Geographers, 67, September 1977, p. 426. By permission of the Association of American Geographers.)

Figure 10.3 World distribution of Ultisols.

Figure 10.4 Ultisols of the United States. (From Philip J. Gersmehl, "Soil Taxonomy and Mapping," *Annals of the Association of American Geographers,* 67, September 1977, p. 426. By permission of the Association of American Geographers.)

Figure 10.5 Water budget of Boromo, Burkina Faso (based on normal data).

Figure 10.6 Water budget of Fort Nelson, British Colombia (based on normal data).

Figure 10.7 Water budget of Atlanta, Georgia (based on normal data).

Figure 10.8 Suborders and Great Groups of the Soil Order Alfisol.

Figure 10.9 Suborders and Great Groups of the Soil Order Ultisol.

Figure 10.10 Soils of Montgomery County, Alabama (Unpublished manuscript of classroom exercises, courtesy of Neil E. Salisbury).

Figure 10.11 Corn yields with nitrogen applications under varied moisture conditions.

Figure 11.1 World distribution of Oxisols.

Figure 11.2 Water budget of Uapes, Brazil (based on normal data).

Figure 11.3 Structure of a mature tropical rain forest.

Figure 11.4 Suborders and Great Groups of the Soil Order Oxisol.

Figure 11.5 World distribution of shifting cultivators.

Figure 11.6 The field patterns of a shifting agriculturist.

Figure 12.1 Histosols of the United States. (From Philip J. Gersmehl, "Soil Taxonomy and Mapping," *Annals of the Association of American Geographers,* 67, September 1977, p. 427. By permission of the Association of American Geographers.)

Figure 12.2 Suborders and Great Groups of the Soil Order Histosol.

Figure AI.1 Structure of the silica tetrahedron.

Figure AI.2 Structure of the alumina octahedron.

Figure AI.3 The clay mineral kaolinite.

Preface: Second Edition

The *Geography of Soils* was first published to meet an immediate need: to provide a concise soils textbook for liberal arts students interested in environmental studies. The first edition was written at a time when few published materials on soil were available for individuals without a background in the natural sciences, and none of these addressed the *Soil Taxonomy*. This revision builds on the strengths of its predecessor, is updated, clarifies concepts that students have experienced difficulty in mastering, and contains new sections. The new sections include an optional discussion of clay mineral structures and a more thorough treatment of soilscapes and suborders within a regional context. It is the authors' hope that our colleagues will find a familiar, yet improved, textbook that meets their needs.

The authors wish to express thanks to their many colleagues and students whose numerous discussions have resulted in clarifying difficult concepts. Special appreciation goes to Kenneth Burrows, Gail Gibson, and Russell Kologiski for their comments on various parts of the manuscript, and to Philip Gershmel, the Association of American Geographers, and Neil Salisbury for permission to reproduce the dot distribution maps and block diagrams found in Chapters 7 through 12.

D. Steila

T. Pond

Preface:
First Edition

Throughout the United States there has been a rapidly developing interest in environmental quality and the recognition of the important role which soil serves as one of man's primary resources. As a consequence many liberal arts students are now including an introductory soil science course in their curriculums. In the process of teaching such a course— *Geography of Soils*—to students with limited training in the natural sciences, it became apparent that a need existed for a textbook written at a level comprehensible to the non-specialist, but one that was not superficial in scope. It is my sincere hope that I have approached meeting these objectives.

Many individuals have assisted in the development of this book. Special recognition should be given to Dr. Jack Blok (Director, ECU Cartographic Laboratory), who personally directed the production of all graphic renderings, and to the East Carolina University Staff cartographers Robert Corbo and Steven Moore. Professors Ronald Swager and Vernon Smith provided assistance when schedules were rushed. Andy Goodwin read the chapters and made valuable contributions to each one. Georgia Arend was a research aide and typed the final manuscript. Fellow colleagues and graduate students at the University of Georgia, University of Arizona, and East Carolina University provided suggestions for improvement. Most important in the production of this manuscript, however, has been my family, who gave constant encouragement and were understanding when research and writing demanded much of my time.

D. Steila

Introduction

Never in mankind's history has soil been so central to human survival. As a rapidly growing population generates an increased demand for the basic needs of food, clothing, and shelter, so we become more aware of the importance of soil resources to supply them.

Human history is a record of the environmental factors that governed the unpredictable access to or deprivation of food. Artifacts dating back to human origins imply hominoid populations were organized into small family groups that roved within limited tracts of land, gathering and sharing what they picked from nature's bounty. Food supply and population size increased or diminished along with climatic and weather variations.

Eventually people learned to control their environment to some extent—domesticating plants and animals, controlling the flow of water to land and crops, producing surplus food—and to form civilizations. Technologic advances leading to increased food production led to rapid population increases. A million years passed before human population became the size it attained in 1650; 200 years later (in 1850), its size had doubled. Presently, it is doubling in only forty years; and if recent trends continue, a world population of more than 6 billion by year 2000 is predicted.

Our ability to meet food requirements of future populations, it is becoming increasing clear, ultimately depends on the soils of the continental landmasses.

Soil responds to mankind's nurturing, such as fertilization and irrigation, with increased plant yields. When abused, as when irresponsible land management leads to excessive erosion, soil productivity diminishes and may even cease to support plant life. Recent statistics on world grain yields (Brown, 1981) indicate that per capita production has leveled off and may be declining. This is sobering news for a species that has lived on the premise that technology can solve all things. It should also heighten an awareness of the soil's importance to human survival. Next to air and water, soil is probably man's most fundamental earth resource. Yet each year, in the United States alone, more than 1 million hectares of rural land is lost to nonfarm purposes, such as buildings, airport runways, and roads (Soil Conservation Service, 1980).

For the security and well-being of future generations, it is imperative that immediate attention be directed toward understanding and protecting our soil resources.

Soil serves as an anchorage for plants and as their nutrient reservoir. Both organic and inorganic in composition, soil is the loose surface material of earth wherein many complex biological, chemical, and physical processes take place. Stated most simply, soil is made up of four components: (1) organic matter, (2) inorganic material, (3) water, and (4) air. Each of these components varies from site to site in both character and amount. This variation is due to the interaction of climate and vegetation with inorganic materials on different geomorphic surfaces. When these interactions occur within relatively homogeneous areas, the soils of the region have certain characteristics in common. However, smaller scale environmental differences within a region can additionally impart a unique character to individual soil units. It is this uniqueness that poses a challenge to the understanding and wise management of soil resources.

The Origin and Significance of the Soil's Inorganic Constituents

1

MINERALS AND PARENT MATERIAL

A soil is composed of approximately 45 percent mineral matter. This mineral portion forms the basic framework of the soil, serving as both an anchorage and a nutrient reservoir for plants. As such, the origin and characteristics of minerals are of prime concern to the *pedologist* (soil scientist).

Minerals are generally defined as naturally occurring elements or compounds formed by inorganic processes. The compounds may be either crystalline or amorphorous. The *crystalline* minerals possess atoms occurring in a definite order that repeats itself in three dimensions. The *amorphorous* (noncrystalline) minerals do not have a regular, extensively repeated atomic structure. Most of the earth's minerals are variously aggregated, or clustered, into types of rock. For example, granite is a common rock type that is composed of many interlocking crystalline minerals, including quartz, micas, and orthoclase and plagioclase feldspars. While each mineral found in the granite has unique characteristics, every one had its origin in exactly the same way.

As the earth evolved from a molten sphere into a solid globe, cooling took place. That cooling resulted in a reduction in the activity or mobility

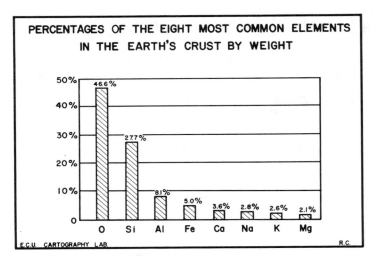

Figure 1.1 Percentages of the common elements in the earth's crust by weight: O = oxygen, Si = silicon, Al = aluminum, Fe = iron, Ca = calcium, Na = sodium, K = potassium, and Mg = magnesium.

of *ions* composing the *magma* (molten rock).[1] With decreased activity, the ions responded to their electrical attractions and became bonded together in fixed positions, producing solid crystalline minerals. The composition of the magma and that of resultant minerals was exactly the same, but in the solid state the ions were arranged according to a definite pattern, or crystalline structure.

All of the naturally occurring elements (special combinations of the protons, neutrons, and electrons) now recognized on earth were present in the molten mass from which the earth was formed. Eight elements dominated the composition of the magma, however, and they now make up over 98.5 percent of the earth's crust by weight (Figure 1.1).

Various combinations of the earth's elements produced a wide variety of minerals. The more important original and secondary minerals are outlined in Table 1.1. Original minerals have been produced dominantly in primary rock types, that is, *igneous*; whereas the secondary minerals are primarily the result of weathering (including the formation of sedimentary rocks). As they occur in their aggregate structure (i.e., rock), original and secondary minerals form the bulk of the earth's outer crust. When

---

[1] An *ion* is an atom, group of atoms, or compound that is electrically charged as a result of the loss or gain of electrons.

Table 1.1   Selected Original and Secondary Minerals

Original minerals	
Name	Formula[a]
Quartz	SiO_2
Microcline	$KAlSi_3 O_8$
Orthoclase	$KAlSi_3 O_8$
Na-plagioclase	$NaAlSi_3 O_8$
Ca-plagioclase	$CaAlSi_3 O_8$
Muscovite	$KAl_3 Si_3 O_{10} (OH)_2$
Biotite	$KAl(Mg \cdot Fe)_3 Si_3 O_{10} (OH)_2$
Horneblende	$Ca_2 Al_2 Mg_2 Fe_3, Si_6 O(OH)_2$
Augite	$Ca_2 (Al \cdot Fe)_4 (Mg \cdot Fe)_4 Si_6 O_{24}$

Secondary minerals	
Name	Formula[a]
Calcite	$CaCO_3$
Dolomite	$CaMg(CO_3)_2$
Gypsum	$CaSO_4 \cdot 2H_2O$
Apatite	$Ca_5(PO_4)_3 \cdot (Cl,F)$
Limonite	$Fe_2O_3 \cdot 3H_2O$
Hematite	Fe_2O_3
Gibbsite	$Al_2O_3 \cdot 3H_2O$
Clay minerals	Al-silicates

[a] For those unfamiliar with chemical formulae for minerals, the following example may be helpful: Quartz has the formula SiO_2. This indicates a chemical bond between one silicon atom (Si) and two oxygen atoms (O_2).

minerals, through subsequent weathering, are released from their bond with adjacent crystals, they become available to supply the major bulk of soil material and the nutrient needs of plants.

PARENT MATERIAL AND ITS TRANSFORMATION

The process leading to the development of a true soil requires considerable time. Numerous alterations must be performed on the surface layer of the earth's crust before a soil capable of supporting plant life develops. Such changes normally begin with the *disintegration* and *decomposition* of exposed rock material. Disintegration signifies a reduction in size of the original material, whereas decomposition refers to the chemical alteration of minerals. Collectively, the processes of disintegration and decomposition are called *weathering*.

Since a soil's characteristics can be strongly dependent upon its framework collection of minerals, the original mineral complex from which a soil is formed (and is still forming) is called the soil's *parent material*.

There are two basic sources of inorganic parent material. One is known as *sedentary* (residual) and the other *transported*. A sedentary parent material is one that is native to the site. Suppose, for example, that a granite outcrop is weathering in place, without any significant removal of material. The soil that is formed will be composed of the residual products of the parent material and will, therefore, be considered sedentary (Figure 1.2). On the other hand, many soils develop from inorganic material that originated somewhere else. A river such as the Mississippi, when overflowing its banks, may deposit sediments that have been transported hundreds, even thousands, of kilometers. Such parent material is not sedentary (Figure 1.2). Besides running water, gravity (down slope movement), glacial ice, waves and offshore currents, and wind may carry inorganic material to a foreign site. Collectively, this type of parent material is said to be transported.

INORGANIC PARENT MATERIAL

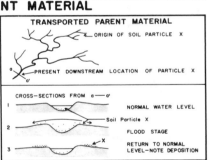

Figure 1.2 Examples of the difference between sedentary and transported parent material.

WEATHERING PROCESSES

When affected by weathering, solid rock may be broken into smaller fragments or ultimately into individual minerals, or fragments of minerals may be modified or altered completely. These processes that either release or alter particles and make them available for soil development fall into two broad groups—*mechanical* and *chemical weathering*.

Mechanical weathering processes involve a physical reduction in the size of rocks and a separating out of the minerals. Several factors are significant:

Unloading. This is a process in which the removal of overlying rocks or sediment reduces pressure on the freshly exposed rock, permitting it to expand and produce cracks and fissures.

Temperature variation. Since each mineral expands and contracts at a different rate when heated or cooled, rocks in environments that experience wide diurnal (daily) temperature ranges develop stresses between the surface mineral bonds—eventually weakening and separating individual crystals. This aspect of weathering occurs only when moisture is present, even if only in minute amounts.

The intense heat of forest and brush fires can raise surface rock temperatures dramatically within a few minutes, causing a steep temperature gradient in the upper rock layer and contributing to severe rock rupture. This process can be likened to a commonly occurring household accident with glassware. If you should quickly pour hot liquid into a cool glass jar, it will usually shatter due to nonuniform heating throughout the thickness of the glass. The same process is operative in rocks experiencing strongly contrasting temperatures.

Frost shattering. This occurs in regions where there are periodic freezes. Water freezing into cracks, crevices, and pore spaces exerts tremendous pressure as it expands, and this can pry apart or even shatter massive rocks. Have you ever put a sealed container with liquid (*e.g.*, a bottle of soda) in the freezer compartment of a refrigerator and forgotten it? Eventually you discovered its contents all over the inside of the freezer. This illustrates on a very small scale the force associated with water expansion due to freezing.

Plants and animals. Plant roots penetrating into cracks and crevices exert a prying effect on rock material, and animals burrowing through soil or dislodging earth fragments also aid the process of physical disintegration. However, the physical influence of plants and animals is minor compared with that of other mechanical weathering processes.

Chemical weathering covers a group of processes in which minerals are altered in composition as well as reduced in size, transforming the original material into something different, that is, converting geologic material (nonsoil) into soil. The main processes involved are hydrolysis, hydration, oxidation, and carbonation. Seldom do they operate individually; rather, they are interrelated. If this were not the case, weathering would be an extremely slow process.

Hydrolysis. This refers to the process in which dissociated H^+ (hydrogen) and OH (hydroxyl) ions of water react with many rock-forming minerals. The effect of water in altering minerals is without doubt *the* dominant chemical weathering activity. The following is an example of the alteration of the mineral orthoclase. When precipitation (H_2O) comes into contact with the mineral orthoclase ($K(AlSi_3O_8)$), the hydrogen ion from the water may disrupt the mineral's crystal structure, producing an aluminosilicic acid ($HAlSi_3O_8$) and a hydroxide (KOH). The aluminosilicic acid, which is unstable, undergoes further change, which may result in the formation of clay minerals through recrystallization.

Hydration. When water combines chemically with other molecules, hydration occurs. Although this process may change the mineral structure, it often affects only the surfaces and edges of mineral grains without modifying their internal structure. An illustration of mineral conversion is the change of anhydrite ($CaSO_4$) in the presence of water (H_2O) to gypsum ($CaSO_4 \cdot 2H_2O$).

Oxidation—Reduction. Oxidation is a process wherein oxygen combines with rock and soil compounds to form oxides. As oxidation takes place, the original material rots and is weakened. Iron, titanium, manganese, copper, and phosphorus are dominant elements that experience oxidation. Normally this process is slow; however, in moist and warm regions (such as the tropics), the process can be rapid. Oxidation takes place in well-aerated materials containing abundant oxygen supplies. A widespread occurrence of the process can be observed on automobiles that exhibit a corrosive substance commonly referred to as *rust*. Similarly, the red color of many soils is associated with iron oxides that coat or stain the soil particles they contact.

Reduction happens where soils are water-logged and become *anaerobic* (oxygen depleted). Under these conditions certain ions, particularly iron and manganese, are converted to ferrous mobile forms capable of being transported into (and sometimes out of) the soil system. If the soil location has an associated watertable possessing external drainage, mobilized ions may be removed from their location and either discharged into surface streams or moved downward to subsurface aquifers. When drainage is impeded and soil water becomes stagnate, ferrous iron may remain in the soil system to form sulfides that can imbue the soil with blue-green and green colors.

Either intra-annual precipitation variation or long-range climatic/ geologic changes can provide suitable opportunities for soil systems to experience alternating reducing and oxidizing environments. Soil in the former state is acidic and releases iron and manganese from parent minerals for transportation and local concentration within the soil. When aeration and oxidizing conditions are reinstated, concentrated iron and manganese form *mottles* (spots or blotches of different color or shades of color interspersed with the dominant soil color). These are evidence of periodic, or cyclic, soil wetness.

Carbonation. Precipitation (H_2O) plus atmospheric carbon dioxide (CO_2) unite to form carbonic acid (H_2CO_3). When in contact with carbonic acid, minerals containing lime, soda, potash, or other basic oxides are altered to carbonates. Should carbonic acid be in contact with limestone ($CaCO_3$), the weathered product will be calcium bicarbonate ($Ca(HCO_3)_2$) in a solution that may be lost from the soil system through subsurface drainage.

Physical and chemical weathering do not occur in exclusion of one another. One group may be more active in a particular area, but both physical and chemical changes usually occur simultaneously. A physical reduction in rock size hastens chemical activity by exposing a greater amount of surface area to attack. This is important since decomposition normally proceeds inwardly from the surface of rock fragments and minerals. A 10-centimeter cube of rock (or mineral) matter has a total surface area of 600 hundred square centimeters. If the cube is evenly broken into eight 5-centimeter cubes, the area exposed to chemical weathering increases to 1,200 hundred square centimeters (Figure 1.3), that is, twice as much.

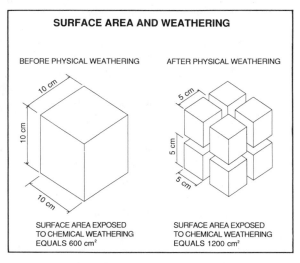

Figure 1.3 Relationship of the size of fragments to total surface area.

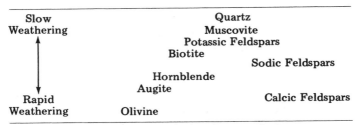

Figure 1.4 Resistance of minerals to weathering.

Another factor affecting weathering processes is that all minerals do not decay at the same rate. Some are more resistant than others and have a longer duration time at the surface. Thus, under specified climatic conditions and types of parent material, resultant soil should contain certain of the original minerals in greater quantity than others. Figure 1.4 illustrates the relative rate at which minerals in common igneous rocks chemically decompose. Resistance to decay reflects, to a large degree, the difference in surface conditions under which minerals experience weathering in relation to conditions existing when the minerals originally solidified. Olivine, when forming from a molten state, crystallizes under high temperatures and high pressures. As a consequence, it tends to be unstable when exposed to low temperatures and pressures at the earth's surface and thus weathers rapidly. Quartz, on the other hand, develops under considerably lower temperatures and pressures—in the late cooling stages of the magma—and it is relatively stable and very resistant to weathering.

With a knowledge of the weathering sequence, it is possible to project the relative amount and type of mineral that will accumulate in the soil through time. Soluble materials such as calcium, magnesium, and sodium are generally removed rapidly in humid regions where moisture is abundant. Under the same climatic conditions, oxides of silicon, iron, and aluminum—residual products of decomposition—are resistant and will most likely accumulate in the soil. A diagram illustrating the differing accumulations of mineral constituents is shown in Figure 1.5. Grouping the *bases*—those minerals containing Calcium(Ca), magnesium(Mg), sodium(Na), and potassium(K)—together, the diagram presents a hypothetical case of residual accumulation of a highly weathered tropical soil and assumes a reduction in volume of the soil column of 50 percent.[2]

---

[2] A base is any molecule or ion that can combine with a protron. Normally within the field of soil science the cations of calcium, magnesium, potassium, and sodium are referred to as *exchangeable bases* or more simply *bases*. This terminology is employed because these four cations are associated with $CaCO_3$, $MgCO_3$, K_2CO_3, and Na_2CO_3, which are commonly occurring non-acidic compounds within the soil.

In a parent material containing equal quantities of the bases, the order of their weathering is as follows:

Calcium (Ca)	Most Rapid
Magnesium (Mg)	↑
Sodium (Na)	↓
Potassium (K)	Least Rapid

Although calcium is most susceptible to weathering, it is often found in the uppermost soil layer in slightly greater amounts than magnesium because (1) plants utilize more calcium than magnesium, therefore returning more calcium to the surface through plant absorption and release upon decomposing; and (2) calcium ions have a stronger affinity for binding to clay minerals than do magnesium ions.

The natural infertility of older tropical (and many other humid land) soils can be understood in the light of the weathering sequence in Figure 1.5. *Fertility* is "the status of a soil with respect to the amount and availability to plants of elements necessary for plant growth" (Soil Science Society of America, 1975). Humanity's food requirements are mainly provided by the grass family of plants, both for cereal grains and as forage for grazing animals. The grass family requires a relatively high availability of bases. Without bases—which we know are highly soluble—grain and forage crops cannot thrive. Anyone familiar with farming practices in

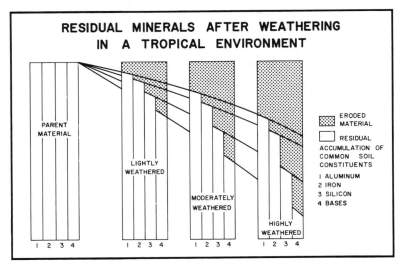

Figure 1.5 Accumulation of mineral constituents as weathering occurs. (The silicon is the mineral portion contained in silicate minerals; it is not quartz.)

much of the eastern United States knows that crushed limestone is frequently applied to fields. The limestone amends infertile soil by replacing calcium lost through soil drainage and plant removal, and it corrects the excessive acidity that limits availability of other plant nutrients.

PRODUCTS OF WEATHERING

As time passes and the parent material undergoes continued weathering, a group of mineral fragments is produced from what was once the more massive original rock complex. Since a soil may be composed of particles of many sizes, the pedologist separates the fragments into groups for identification. The individual types are called *soil separates*. The name of each soil separate is shown in Table 1.2, along with its size range.

Sand grains, when dominant, yield a soil that can be easily worked by the farmer. Consequently, a sandy soil is said to be *light*. The major sand mineral is usually quartz (SiO_2), even though coarser sand may contain rock fragments of varied composition (Figure 1.6). Dominance of quartz tends to make sand fractions of soil chemically inactive. Nutrients released by weathering are quickly leached away in humid climates,[3] since sandy soils are typically highly permeable and have very low water-holding capacity. So, although sandy soils are easily worked, they are limited for crop production by rapid leaching, low nutrient and water-holding capacities, and low natural fertility.

Table 1.2   Soil Separates

Soil Separate[a]	Diameter (mm)	Diameter (in.)
Very coarse sand	2.00—1.00	0.08—0.04
Coarse sand	1.00—0.50	0.04—0.02
Medium sand	0.50—0.25	0.02—0.01
Fine sand	0.25—0.10	0.01—0.004
Very fine sand	0.10—0.05	0.004—0.002
Silt	0.05—0.002	0.002—0.00008
Clay	below 0.002	below 0.00008

[a] Stone, gravel, and organic material are listed separately and generally are not included in the fine earth analysis.

---

[3] *Leaching* is the removal of soluble materials in a solution from the soil.

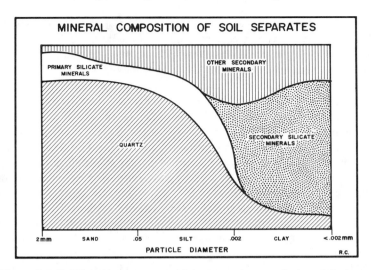

Figure 1.6 Relationship between particle size and types of minerals present.

The chief function of sand is to serve as a framework for the more active soil components. Due to their relatively large size, sand particles provide the greatest degree of space between individual grains, thus promoting the movement of air and the drainage of water within the soil.

Silt, like sand, is composed dominantly of silicate minerals. It differs from sand mainly in its smaller particle size. Consequently silt has a total surface area per unit volume of soil greater than that of sand. In addition, it has a faster weathering rate, releases soluble nutrients for plant growth more readily, and retains more "available" water than does sand.

Clay covers a wide range of substances with varied mineralogical and chemical characteristics. It is defined as mineral particles less than 2 microns in diameter, and generally falls into one of two broad groups—the silicate clays and the iron and aluminum hydrous-oxide clays. The silicate clay group is typical of weathering processes in the middle latitudes, whereas the hydrous-oxide clays characterize soil types found in the tropics.

Most clay crystals resemble minute flakes in which atoms are arranged in a layered structure. The layers are of two forms: *silica sheets* and *alumina sheets.* A silica sheet is composed of tetrahedrons (pyramids), each having an oxygen atom at the vertices and a silicon atom in the interior. Each oxygen atom is equally spaced and equally distant from the silicon atom. By sharing oxygen atoms at the base of the pyramid, tetrahedra combine into hexagonal units. In repeating this pattern, these units form a *lattice* of the clay mineral. An alumina sheet consists of octahedrons, each having a central atom of aluminum equidistant from six oxygen atoms or hydroxyls. The combinations of silica and alumina sheets relate to their stability and provide a basis for clay classification.

The very minute size of clay particles exposes an extremely large surface area (per unit volume) upon which chemical activity can take place. Unlike the more sterile sand separates, clay is an active portion of the soil—holding and exchanging ions.

Clay minerals form by two principal processes. They can be formed directly by the alteration and reduction of parent material or indirectly by synthesis from weathering products. *Alteration* occurs when chemical weathering removes certain soluble components and substitutes others. *Recrystallization* is a complete change in structure of the original minerals; for example, *kaolinite,* the simplest of the clay minerals, may develop from solutions containing soluble aluminum and silicon. Figure 1.7 illustrates the weathering sequence associated with production of clay minerals and also identifies various weathering stages. Chlorite and hydrous micas represent youthful stages, and kaolin the old-age stage, of silicate weathering. (For a more thorough discussion of clay structure, the reader should consult Appendix I.)

In contrast to sandy soils, which are considered light, clay soils are referred to as *heavy.* The very finely divided particles, packed closely together, create a dense soil of high potential plasticity and particle cohesion. When plasticity is high, dry soil may become hard and cloddy, and wet soil may be a sticky mass that can be molded or deformed by relatively moderate pressure. Plowing such a soil when wet can further reduce pore space, in some cases making the soil impervious to water and air movement. When these soils become dry, they are very hard and dense. Such soils are then said to be *puddled.*

Although soil separates—sand, silt, and clay—are discussed here as individual entities, *texture* refers to the relative proportions of separates in a given soil.

Texture is important for several reasons: (1) It influences the ease with which plant roots penetrate into soil; (2) it affects aeration (exposure to air); (3) it influences moisture storage capacity; and (4) it establishes the rate of chemical reactions within the soil.

For uniformity in soils description, the U.S. Department of Agriculture uses a classification based upon the percentages of various soil separates in specific combinations. Each soil class is defined according to the relative proportions of the soil separates it contains. The USDA soil

GENESIS OF SILICATE CLAYS

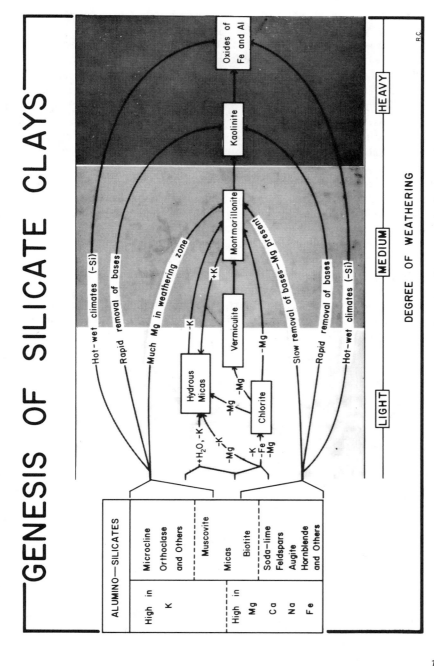

Figure 1.7 Conditions in which the various silicate clays and oxides of iron and aluminum may form.

15

texture triangle (Figure 1.8) provides specific, separate ranges for each soil class, along with its appropriate name.

Most of the terms used have been discussed, except for *loam* and *loamy*. Note on the diagram that the loam soil class has substantial amounts of all three soil separates. Indeed, it represents a "blend" of sand, silt, and clay. When the word *loamy* is used as a prefix (*e.g.*, loamy sand), it suggests an "intergrade" between the pure sand and sandy loam soil texture classes.

Determination of soil texture class is either done in a laboratory or in the field. Laboratory techniques involve (1) passing a soil sample through nested sieves that separate the larger particles by size and then (2) separating silt and clay components via monitoring the rate at which they settle to the bottom of a cylinder containing water and a dispersing chemical. Obviously, the heaviest particles settle first and the smallest last. Measuring the solution's density at specified time intervals permits an analyst to determine the proportion of solids of a given size.

Field techniques are much quicker than laboratory procedures. They do not provide the specificity of laboratory analysis but do allow for meaningful soil class determinations. The procedure is simple. A moist soil sample is mixed in the hand. The next step is to determine if it will form a cast in the hand or how it behaves when squeezed between the forefinger and thumb.

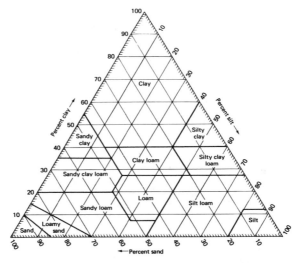

Figure 1.8 Chart showing the percentage of clay (below 0.002 mm), silt (0.002 mm to 0.05 mm), and sand (0.05 mm to 2 mm) in the basic soil textural classes.

Provided below is a field guide to aid you in classifying the most common soil texture classes:

Sand. Sand is loose and single-grained. The individual grains can readily be seen or felt. Squeezed in the hand when dry, it will fall apart when pressure is released. Squeezed when moist, it will form a cast that will crumble when touched.

Sandy loam. A sandy loam is a soil containing much sand, but with enough silt and clay to be somewhat coherent. The individual sand grains can readily be seen and felt. Squeezed when dry, it will form a cast that readily falls apart; squeezed when moist, it will form a cast that will bear careful handling without breaking.

Sands and sandy loams are classed as coarse, medium, fine, or very fine, depending on the proportion of the different sized particles that are present.

Loam. Loams are soils having a relatively even mixture of sand, silt, and clay. They are mellow with a somewhat gritty feel, yet fairly smooth and slightly plastic. Squeezed when dry, they will form casts that will bear careful handling. The cast formed by squeezing moist soil can be handled quite freely without breaking.

Silt loam. A silt loam has a moderate amount of the fine grades of sand and only a small amount of clay, more than half of the particles being of the size called silt. When dry, it may appear quite cloddy, but the lumps can be readily broken; when pulverized, it feels soft and floury. When wet, the soil readily runs together and puddles. Either dry or moist, it will form casts that can be freely handled without breaking. If squeezed between thumb and finger, it will not "ribbon" but will give a broken appearance.

Clay loam. Clay loams are fine-textured soils that break into clods or lumps that are hard when dry. When the moist soil is pinched between the thumb, it forms a thin "ribbon" that will break readily, barely sustaining its own weight. The moist soil is plastic and forms a cast that will bear much handling. When kneaded in the hand, it does not crumble readily, but tends to work into a heavy, compact mass.

Clay. Clay is fine-textured soil that forms very hard lumps or clods when dry and is quite plastic and usually sticky when wet. When the moist soil is pinched between the thumb and fingers, it forms a long, flexible "ribbon."

SOIL STRUCTURE

Structure is the arrangement of soil particles (sand, silt, and clay) into larger secondary units called *peds* or *aggregates.* Soil structure is the result of many factors: the chemical nature of the clay, the amounts of clay and organic material in the soil, the composition of soil microorganisms, wetting and drying, freezing and thawing, and cultivation.

Like texture, structure is an important physical characteristic of the soil. Although most soils are dominantly inorganic in composition, organic matter is extremely significant in the development of certain types of structure. Some soils, such as those composed entirely of sand, may show no structural arrangement due to the lack of binding elements. Other soils may have peds of varying shapes and sizes.

Soil structure is classified according to *grade, class,* and *type.* In the North Carolina Piedmont, for example, the Cecil soil series has a structure classification of *moderate-fine-subangular blocky,* where *moderate* is the grade, *fine* is the class, and *subangular blocky* is the type (shape of the structure).[4] Class and type designations are provided in Figure 1.9. *Grade* refers to structure strength and is dependent upon moisture content. If peds

SOIL STRUCTURE: TYPES AND CLASSES

		PLATELIKE	PRISMLIKE	BLOCKLIKE	
Type of Ped		PLATY	PRISMATIC OR COLUMNAR	BLOCKY Angular or Subangular	SPHEROIDAL Granular or Crumb
Ped Size Classes	VERY FINE OR VERY THIN	1 mm	10 mm	5 mm	1 mm
	FINE OR THIN	1 – 2 mm	10 – 20 mm	5 – 10 mm	1 – 2 mm
	MEDIUM	2 – 5 mm	20 – 50 mm	10 – 20 mm	2 – 5 mm
	COARSE OR THICK	5 – 10 mm	50 – 100 mm	20 – 50 mm	5 – 10 mm
	VERY COARSE OR VERY THICK	10 mm	100 mm	50 mm	10 mm

Figure 1.9 The common manner in which soil separates are arranged into structural units and their size class.

---

[4] *Soil Series.* The basic unit of soil classification consisting of soils that are essentially alike in all major characteristics except the texture of surface soil separates.

or aggregates are observable in place but cannot be removed without being destroyed, the structure's grade is *weak*; it is *moderate* when peds can be removed for hand examination, and it is *strong* when they are rigid and durable in the hand.

One theory to explain aggregation, or soil particle clustering, concerns the electrical charge of colloidal particles.[5] Water molecules are *dipolar* in charge and can be firmly attached to colloidal nuclei; in fact, may serve as a link between two colloids.[6] As water evaporates from soil, the combining link is effectively shortened and draws together colloidal particles and larger soil grains to which they are attached. This process effectively clusters colloids and larger soil particles. As evaporation continues and colloidal material becomes further dehydrated, soil particles may stick or cement into an aggregate.

Other significant factors include the folowing:

1. The *mycelial* (sort of cobwebby filaments) growth of microorganisms that serves as a mini root system, binding particles together. Microbial gums may act as cements and can be the most important part of aggregate formation.

2. Water that has frozen in the pore space of soil exerts pressure on surrounding soil particles as ice crystals expand. Pressing together stimulates the clustering of colloids and other soil particles. Furthermore, as moisture is withdrawn from the surrounding soil, colloidal material is dehydrated, resulting in the cementing of particles in contact with one another. Thawing provides increased pore space in the areas previously occupied by ice crystals.

Structure is important because it changes the influence of soil texture on such factors as infiltration of moisture, aeration, ability of plant roots to penetrate the soil, nutrient supplies, and the soil's resistance to erosion.

MICELLE ACTIVITY AND PLANT NUTRIENT AVAILABILITY

Earlier, the importance of the clay fraction in soil separates was stressed. These particles, less than 2 microns in diameter, are chemically active and represent the plant's main reservoir for nutrients. Pedologists also recognize *colloids* within this category. Colloids can be either organic or inorganic, but they are less than 1 micron in diameter. For now we are concerned only with inorganic colloids, also known as *micelles* (microcell).

---

[5] *Colloid* refers to organic and inorganic matter having a very small particle size and a high surface area per unit mass.

[6] *Dipolar* means having two poles as a result of the separation of electric charge. A dipolar molecule orients in an electric field.

Micelles, minute silicate-clay colloids, normally have negative charges. Thus, they have a capacity to attract and hold positively charged ions, called cations, that are released during the weathering of the parent material or added in fertilizers.[7] Table 1.3 summarizes products of some weathered minerals and also lists the associated ions released during the weathering processes.

Soilwater usually carries the cations to micelles for absorption.[8] As water percolates through soil, it can transport dissociated ions and bring them in contact with the micelle (Figure 1.10). The absorption of cations and their subsequent release is known as *cation exchange*. Each soil has a unique *cation-exchange capacity* (CEC),[9] depending on the existing amount of exchangeable ions. The quantity of exchangeable ions, in turn, is dependent upon (1) the type of clay mineral, (2) the percentage of clay, and (3) the amount of organic colloids. In addition to a unique CEC, each soil type is characterized by a specific proportion of its cations being bases. *Percentage base saturation* (PBS) provides information on the relative amount of exchangeable bases (for example, calcium, magnesium, sodium, and potassium) to a soil's total exchange capacity. In quantitative terms,

$$\text{Percentage base saturation (PBS)} = \frac{\text{Exchangeable bases (milliequivalents)} \times 100}{\text{CEC (milliequivalents)}}$$

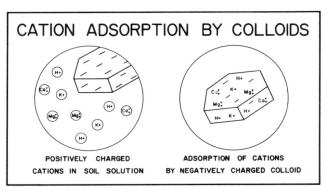

CATION ADSORPTION BY COLLOIDS

POSITIVELY CHARGED
CATIONS IN SOIL SOLUTION

ADSORPTION OF CATIONS
BY NEGATIVELY CHARGED COLLOID

Figure 1.10 Diagram of a clay colloid (micelle) with its sheetlike morphology, numerous negative charges, and swarm of absorbed cations.

---

[7] *Cations* have a positive charge, as opposed to *anions*, which have a negative charge.

[8] *Adsorption* refers to the adhesion of dissociated ions to soil colloids, whereas *absorption* involves ion assimilation.

[9] *Cation-exchange capacity* (CEC) is the sum total of exchangeable cations that a soil can absorb, expressed in milliequivalents per 100 grams of soil. An *equivalent* represents a quantity that is chemically equal to 1 gram of hydrogen. The number of hydrogen atoms in an equivalent equals 6.02×10^{23}. A *milliequivalent* is 6.02×10^{20} (0.001 gram of hydrogen).

Table 1.3   Weathering Products of Selected Common Minerals

Original mineral	Residual mineral products of weathering	Released ions
Amphibole	Clay minerals, limonite, hematite	K, Ca, Mg, Na
Biotite	Clay minerals, limonite, hematite	K, Mg
Calcite	None for mineral; from limestones some quartz, clay minerals, and hematite as impurities.	
Chlorite	Clay minerals	Mg, Fe^{++}
Clay minerals	Under high moisture in tropics may develop bauxite	SiO$_2$
Dolomite	None from mineral; from dolomite rock some quartz, clay minerals and hematite as impurities.	Ca, Mg, HCO$_3$
Garnet	Garnet	
Gypsum	None for mineral; from gypsiferous shale or sandstone, clay minerals and quartz.	Ca, SO$_4$
Hematite	Hematite	
Limonite	Limonite	
Muscovite	Muscovite tends to remain, eventually alters to clay minerals and quartz.	K, SiO$_2$
Olivine	Clay minerals, limonite, hematite	Mg, Fe^{++}
Orthoclase & Microcline	Clay minerals, quartz	K, SiO$_2$
Plagioclase	Clay minerals	Na, Ca
Pyroxene	Clay minerals, limonite, hematite	Mg, Fe^{++}, Ca
Pyrite & Marcasite	Limonite, hematite	Fe^{++}, SO$_4$
Quartz	Quartz	Some SiO$_2$
Serpentine	Serpentine	
Talc	Talc	

Base saturation of soils in arid climates—where moisture is low and soluble bases are not leached out—tends to be high.[9] In humid climates, PBS is generally low because greater amounts of water percolating through the soil remove its bases. Importance of percentage base saturation is directly related to the availability of the bases for meeting nutrient needs of plants; that is, as an index of fertility.

One particular cation, hydrogen (H^+), in appreciable quantity on the surfaces of micelles produces soil acidity. An absence of H^+ and a large

---

[9] *Leaching* is the removal of soluble materials in a solution from the soil.

amount of basic cations produce an alkaline situation. The degree to which the H^+ ion is present in soil provides a measure of nutrient availability, because when hydrogen cations dominate the micelles, the H^+ displaces bases and subjects them to removal by downward-moving soil water. The dominance of H^+ in some soils is related to the chemical force with which ions are held. In general, the following order represents the strength of ion attachment to colloids and the order of ion accumulation in humid area soils:

$$H > Ca > Mg > K > Na$$

Pedologists have devised a measure of H^+ concentration known as pH. Most soils vary in pH from about 4 to 10, and are distinguished from one another by their positions on the pH scale (Figure 1.11). Each unit change on the pH scale represents a tenfold change in concentration of H^+. A pH value of 7 indicates neutrality, that is, H^+ and OH^- ions are in equal concentrations.[10]

Knowing soil pH is extremely important for ensuring proper soil management. The lower the pH, the greater the acidity and decreased availability of base nutrients. In such a case, an application of crushed

---

[10] The neutral point (pH of 7) is established for the concentration of H^+ ions in pure water at a temperature of $24°C$ ($75°F$). This amounts to 1.0×10^{-7} grams of H^+ ions per liter of water and is also equal to the number of OH^- ions present. Since very small numbers, such as 1.0×10^{-7}, or 0.000,000,1, are difficult to visualize, the pH expression is given as the negative logarithm of the hydrogen activity of a soil.

pH Normality of H^+	Acidity Exponential expression
0 1.0	1.0×10^{0}
1 0.1	1.0×10^{-1}
2 0.01	1.0×10^{-2}
3 0.001	1.0×10^{-3}
4 0.000,1	1.0×10^{-4}
5 0.000,01	1.0×10^{-5}
6 0.000,001	1.0×10^{-6}
7 0.000,000,1	1.0×10^{-7}
8 0.000,000,01	1.0×10^{-8}
9 0.000,000,001	1.0×10^{-9}
10 0.000,000,000,1	1.0×10^{-10}
11 0.000,000,000,01	1.0×10^{-11}
12 0.000,000,000,001	1.0×10^{-12}
13 0.000,000,000,000,1	1.0×10^{-13}
14 0.000,000,000,000,01	1.0×10^{-14}

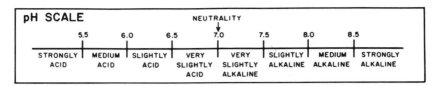

Figure 1.11 The pH scale to measure the concentration of hydrogen cations (H^+) in the soil.

limestone may be necessary to supply the soil with the base calcium (Ca) and to raise the pH to a desirable level.

SOIL COLOR

Color is a necessary part of any discussion concerning parent material as an influence on soil characteristics. Soil color, one of its most readily observable features, results from various chemical weathering processes acting upon the parent material, coupled with organic material. (Organic material, not discussed here, can be a dominant coloring agent in the soil—the carbon content gives it a dark color.)

The most important inorganic coloring agent is iron (Fe). Iron compounds in the parent material produce a variety of colors:

1. Under poor drainage conditions, where oxygen is deficient, chemically reduced iron yields gray to bluish-gray colors.

2. Sites of good aeration and drainage lead to the oxidation of iron, producing red colors.

3. Where the soil is moist a great deal of the time, the iron may be hydrated as well as oxidized, resulting in a yellow color.

Other soils, especially in arid regions, may have accumulations of mineral salts. These soils often show a whitish surface mineral encrustation.

Obviously, color provides a clue to drainage qualities and the sort of parent material from which soil has evolved. But color is only a clue; it is not the complete answer to soil characteristics and processes. A red shale or sandstone, when weathered, may yield a red-colored soil, even though the oxidation of iron is not the primary process.

The Organic Fraction of the Soil

2

The organic portion of the soil is extremely varied and complex, acting as both a substance and as an agent in decomposition. Organic matter averages only about 5 percent by weight in an ideal loam, but its role in influencing soil properties and as fundamental to the biochemical activities taking place within the soil is of great importance.

SOIL ORGANISMS

From microscopic, single-celled organisms to large burrowing animals, the soil is replete with life. It is anything but an inert substance, as an examination of the often unseen plant and animal inhabitants would prove. Within a single cubic centimeter of fertile topsoil, over a billion bacteria may be present; under normal conditions in a single gram of soil, protozoa may number as many as a million and earthworms may exceed 600,000 per hectare. Yet, these life forms represent only a few of the soil's many "tenants."

Microorganisms by far outnumber the larger plants and animals of the soil. The reason is that the soil provides them with a favorable habitat, plus food and water. Unlike higher plants, which can change energy into food through photosynthesis, microorganisms primarily obtain their energy from the tissue of higher plants. In the process of decomposing plant tissue (e.g., organic residue), the microbes are capable not only of obtaining energy, but also of simultaneously releasing back to the soil the many nutrients existing as complex organic compounds within plants. This is an extremely important natural process that permits a recycling of plant nutrients. If a given stand of plants absorbed most of the available nutrients in the soil, future plant growth would be limited as a result of diminished food supply. Microorganisms, however, through their feeding habits, make previously absorbed plant nutrients once again available as soluble inorganic compounds. Some of the more important soil organisms are shown in Figure 2.1 and are briefly described in the following pages.

SOIL FLORA

Plant life in the soil varies both in form and function. The size of plants ranges from the large roots of trees to forms that are less than a

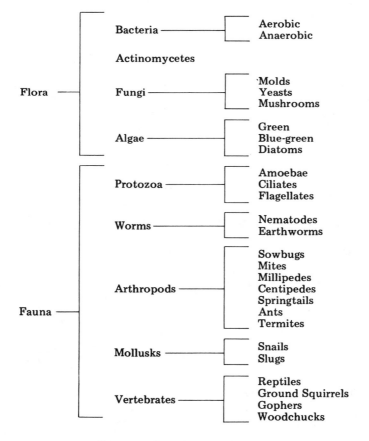

Figure 2.1 Some important soil organisms.

micron in length, and their number is highly dependent upon environmental conditions. The most numerous, by far, are the smallest and most simply structured floral groups (microflora), which can be either increased or limited in number by nutrient supply, temperature, moisture, or pH level of the soil environment.

Certain types of microbes obtain energy by oxidizing simple inorganic substances. Most, however, derive their primary source of energy and their essential elements from organic materials in the soil. While they may carry on decomposition over a wide temperature range, warmer regions usually encourage greater microbial activity. The maximum efficiency in decomposition is estimated at approximately 35°C (95°F). Moisture and aeration, closely associated, are also critical to the type and number of flora present. For *aerobic* (free oxygen-dependent) populations, the most desirable soil moisture is approximately 50 to 70 percent of "field capacity." Water logging and an atmospheric gas deficiency can create a habitat suitable only to *anaerobic* (not dependent on free oxygen) life forms. Also, certain soil organisms cannot tolerate acid or alkaline extremes. Acid soils are favorable for growth of fungi but unfavorable for development of legume bacteria, nitrifying organisms, and actinomycetes. Each type of soil occupant has a unique set of habitat requirements that enables it to perform specific and necessary tasks, normally those of decomposing organic residues and recycling mineral nutrients.

Bacteria. Bacteria are the simplest structured and most numerous life forms existing in soil. Functioning as decomposing agents and—in many instances—nitrogen fixers, they might be considered as one of the most important groups of soil inhabitants. As single-celled organisms they are able to achieve dense population in rather short periods by elongating and dividing into two parts. This microfloral group can be subdivided into aerobic and anaerobic forms. Another subdivision can be made according to their energy source. *Autotrophic* bacteria obtain energy by oxidizing inorganic material; *heterotrophic* bacteria receive energy by decomposing organic materials.

The average sized bacteria is 1 micron long by 0.5 micron in diameter and usually functions best under oxygen and moisture conditions that are considered most desirable for higher plants. While capable of surviving under a wide range of temperatures, their activity is greatest between 21°C (70°F) and 38°C (100°F). Certain bacteria groups can function at low pH levels, whereas others perform their work more efficiently at high levels. Most, however, function best at pH values near neutrality—between 6 to 8—and when there are sufficient amounts of exchangeable calcium in the soil.

In addition to decomposing organic residue, probably the most significant role of soil bacteria concerns their efficiency in immobilizing nitrogen to a form in which higher plant life can assimilate it. Nitrogen is essential for a good rate of plant growth and respiration. When it is not present in sufficient quantity, a plant's ability to utilize basic nutrients is severely limited, its growth is stunted, and its leaves lose their deep green color, becoming pale and yellow due to a loss of chlorophyll.

Nitrogen fixation (immobilization), although not solely confined to bacteria, involves the efforts of numerous members of this microfloral group—the *ammonifiers* (ammonia producers) and the *nitrifiers* (nitrate and nitrite producers). Figure 2.2 illustrates the interrelationships and functions of the more important types of soil bacteria.

Most soil bacteria require nitrogen either in mineral form, such as ammonium salts and nitrate, or as organic nitrogen compounds, such as plant and animal proteins. Very few microflorals are capable of utilizing nitrogen gas in its atmospheric free state. One of the few soil bacteria that can do so is the genus *Rhizobium*. Rhizobium are known as *symbiotic* bacteria; that is, they live in a mutually beneficial relationship with other organisms—normally with the leguminous family of plants. These bacteria infect the root system of plants, develop nodules as they grow, and in a parasitic fashion derive food and minerals from their hosts. In turn, *Rhizobium* are able to utilize atmospheric nitrogen and also to transfer it to the plant.

Among the *nonsymbiotic* (independently existing) bacteria, the most significant in terms of assimilating free nitrogen into their cellular structure

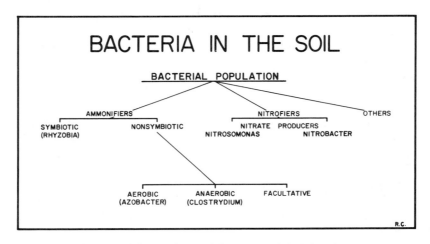

Figure 2.2 Selected bacterial groups and their functions.

are genus *Azobacter* and genus *Clostridia,* both heterotrophic. The specific members or species dominating the *rhizosphere* (root zone) are dependent upon habitat status. Under conditions of good aeration, high organic content, neutral or alkaline reaction (pH $\geqslant$ 6.0), and available calcium, the aerobic *azobacter* are most numerous. (Soil pH seems to be their most restrictive growth factor.) *Clostridia,* on the other hand, are anaerobic— perhaps even *facultative,* meaning they can exist in either an aerobic or anaerobic state. They can withstand highly acidic conditions and are spread over a wide geographic area. Since even an aerated soil may have either water pockets or saturated pore spaces, *Azobacter* and *Clostridia* can operate within the same soil simultaneously.

Bacteria's role in making nitrogen available to higher plants can be better appreciated when it is realized that most of the nitrogen stored within the soil is associated with organic matter. Soil organic matter generally contains 5 to 6 percent nitrogen, most of which is immobilized in protein form—primarily as amino compounds—and is unavailable to higher plants. In the process of digesting organic matter to obtain energy for growth, heterotrophic bacteria set free ammonia (NH_4) as a byproduct. This process is known as *ammonification.* The released NH_4 may then (1) be reappropriated by the bacteria, (2) be utilized by other microorganisms such as mycorrhizal fungi, (3) be made available to higher plants, (4) become fixed by clay minerals and organic matter, or (5) become nitrified.

The *nitrification* process is the work of another group of special purpose autotrophic bacteria known as *nitrobacteria.* Within this group are *nitrosomonas,* which convert ammonia into *nitrites,* that are in turn oxidized to *nitrates* by *nitrobacter.* These end products are in stable forms and can be readily assimilated by plants.

Nitrogen fixation and recycling are possibly the most important roles of soil bacteria, but other bacterial functions are also of major significance. In the process of digesting plant residues, the microbial population temporarily immobilizes carbon, energy, and numerous nutrients. These nutrients once again become available for mineralization and plant use upon the death and decomposition of the microbials. The net effect is not only the recycling of nitrogen, but also of sulphur, phosphorus, calcium, magnesium, potassium, and other nutrients. Hence, these minute and unseen plants, at times numbering in the millions per gram of soil, function as efficient converters of once used materials into forms that again are available for higher plant life.

Actinomycetes. Only bacteria are more numerous in the soil than these organisms, which display morphological features of both bacteria and fungi. Actinomycetes are one-celled organisms of approximately the same characteristics in cross section as bacteria, but they also possess long, threadlike, branched filaments. Because of their appearance they are sometimes referred to as ray fungi. Their total population is approximately 10 to 25 percent less than bacteria; but due to their larger size and filaments, their gross organic weight in a unit volume of soil is about the same as bacteria.

The primary activity of actinomycetes centers on the decomposition and *humification* of organic residues.[1] They are extremely active (and most numerous) in organic material that is in late stages of decay and are especially important in the further breakdown of the more resistant organic compounds. These microbes can function in soils of low moisture and have the greatest moisture-range tolerance of any of the microflorals, as long as available calcium is not restrictive. Actinomycetes prefer soil ranging in reaction from 6.0 to 7.5 and are very sensitive to acidity. Normally they do not exist in soils at a pH of 5.0 and lower.

In addition to their decomposition role, certain species of actinomycetes produce antibiotic substances of value as medicines, while others cause potato scab and pox in sweet potatoes. The devasting potato famine in Ireland during the 1800s was likely due to soil infestation by actinomycetes.

Fungi. The fungi floral class is generally subdivided into three major groups: mushrooms, molds, and yeasts. (The last group is very limited and not of importance in most of the earth's soils.) Fungi are heterotrophic plants, ranging in size from microscopic molds to large and complex fruiting mushrooms and bracket fungi. Of the bacteria, actinomycete, and fungi population, fungi account for probably no more than 1 percent of the total, even though in actual amount of cell substance the combined weight of the fungi group may equal or exceed that of bacteria. This results from the group's high degree of efficiency in transforming the substances which they attack into their tissues. Estimates are that as much as 50 percent of organic wastes decomposed by molds may become fungi tissue.

Fungi perform several important functions, including (1) decomposition of resistant organic materials, (2) extension of the effective rooting system of some plants, and (3) production of antibiotic substances. Molds and mushrooms are capable of vigorous activity in soils, ranging from acid through alkaline reactions. Thus, in highly acid soils where growth of bacteria and actinomycetes is restricted, fungi may become the dominant decomposer of organic residues. The persistence of these plants enables them to decompose even very resistant lignin.

---

[1] *Humification* is a process or condition of decay in which plant or animal remains are so thoroughly decomposed that their initial structures or shapes can no longer be recognized.

Some fungi can extend the effective feeding area to specific plants in a "fungus root" association known as *mycorrhizae*. These associations are most common in forested areas and appear to be of mutual benefit to the plant and the fungi. Since *mycelial* threads of fungi have a tremendous surface absorbing area, the fungi in contact with a root system act as a root extension, increasing the soil contact area several hundredfold and adding to the food supplies of the plant.[2] One gram of soil may contain 10 to 100 meters of mold filament. The fungi, on the other hand, are able to meet their food requirements by absorbing carbohydrates from plant roots. This process is necessary for the growth of certain types of forest species in nutrient-deficient soils, where a mature forest association could not develop without mycorrhizae.

Certain fungi groups have received much attention due to their ability to produce antibiotic substances. The best known is *Penicillium*. Soil research in the Soviet Union has revealed that species of *Penicillium* amount to approximately half of the total fungi population in most Russian soils, about 30 percent in semiarid and 21 percent in desert soils.

Fungi are capable of vigorous growth under a wide range of environmental conditions. The most restrictive factor affecting their growth is oxygen availability.

Algae. Algae are simple structured, chlorophyll-bearing organisms that are capable of carrying on photosynthesis in a manner comparable to higher plants. Although they are the highest form of soil microflora, their total number is only a small fraction of the microbial population. Some algae forms exist well beneath the soil surface by using plant residues or soil organic matter for their food supply. Most species, however, live close to the surface in the presence of light, which supplies the energy needed to combine carbon dioxide and water for the manufacture of carbohydrates.

The common forms in which algae occur in soil are (1) blue-green, (2) green, and (3) diatoms, the least numerous. Blue-green algae can fix nitrogen to form protein, especially in water-logged soils (such as rice paddies).

Algae, which are found worldwide, are among the first occupants of exposed rock material, aiding in the formation of an initial organic layer in the process of soil development. Nevertheless, their role in contributing organic matter and altering the characteristics of mature soils seems to be of minor importance.

---

[2] *Mycelia* are the threadlike bodies of simple organisms, such as the common bread mold.

SOIL FAUNA

The soil's animal population is no less varied in form and function than are the flora members. Here, too, the most numerous are the most minute. Whether these animals are micro or macro in size, they are all involved in the task of altering certain aspects of soil environment. The larger inhabitants are often quite effective in changing the soil's physical characteristics. Through digging and burrowing they may pulverize, mix, or transport earth materials both vertically and horizontally. In the process, they help increase aeration and drainage and alter soil structure. All soil fauna are capable of translocating organic matter and altering it chemically through digestion, thus hastening decomposition.

In combination, soil animals represent an organic component of the soil. When their life cycles are ended, they become substrate for decompositional activity. One earthworm might not seem to be a significant organic contribution, but as many as 600,000 earthworms may be found per hectare of soil, and their total body weight may amount to more than 1,300 kilograms. Thus, individual size cannot be the sole factor in assessing the impact of the specific soil fauna.

Protozoa. Single-celled protozoa are the most profuse members of the soil animal kingdom, numbering over 1,000,000 per soil gram. Their structure is more complex and their size slightly larger than bacteria.

Protozoa are classified into three broad groups based on morphological features: (1) *amoeba,* (2) *ciliates,* and (3) *flagellates.* As the name suggests, ciliates have *cilia* (hairs), and flagellates possess whiplike appendages known as *flagella.* Flagellates are the most numerous of the protozoa, followed by amoeba and then ciliates.

The protozoa live in aquatic environments—in the films of water surrounding individual soil particles. If food supplies become limited or if the soil dries out, these animals are able to *encyst* (enclose as if in a capsule) and later reactivate when conditions are more favorable. Their feeding habits are varied. Some feed on bacteria and in the process stimulate the growth rate of the nitrogen fixers of that group. Others feast on fungi, algae, or dead organic matter. Protozoa effectively hasten decomposition by chemically altering organic components of the soil through enzymatic digestion.

Worms. Nearly everyone is familiar with the common earthworm. Yet, far more numerous in the soil are microscopic and nonsegmented *nematodes,* also known as *threadworms* or *eelworms.* Several billions of them may be found in the surface layer of one hectare of cropland. Nematodes are classified according to feeding habits. One group feeds on decaying organic material; a second is predatory, feeding on other nematodes, earthworms, bacteria, and protozoa; and a third is parasitic, infesting the roots of higher plants. Nematodes can be both beneficial and harmful. As decomposing agents, they bring about a mixture of mineral and organic matter. When too abundant, however, the parasitic group can cause considerable damage to plants by infesting roots and by increasing susceptibility of plants to attack by other parasites.[3]

Due to their widespread distribution, careful land management practices must be exercised to avoid a concentrated population of nematodes. An example of how difficult this can be is shown in the problem plaguing tobacco farmers in eastern North Carolina. Tobacco is a crop especially susceptible to attack by the "root knot," "meadow," and "stunt" members of the parasitic nematodes. In an effort to reduce their numbers, farmers rotate tobacco with crops that are unfavorable hosts for these parasites or else fumigate the soils. The complexity of the nematode problem is compounded by the fact that among the species and strains are types that differ in crop preferences. A corn and tobacco rotation gives the tobacco protection from certain "root knot" strains that cannot thrive on corn, but corn is a favorable host for the "meadow" and "stunt" members, which will increase in numbers and infest tobacco when it is planted. Hence, although a specific crop rotation may reduce the number of some parasitic species and strains, it can enhance the growth potential for others. Proper rotation, coupled with fumigation, may be necessary in many instances to control the adverse effects of these soil microbes. This involves a significant investment of both time and capital.

Considerably larger in size than a nematode, and probably the best-known soil inhabitant, is the common earthworm. Earthworms vary in size to a considerable extent, some species being as long as 45 centimeters and up to 1.3 centimeters in diameter. Ingesting organic matter for food and inorganic matter by burrowing, earthworms chemically alter soil as these materials pass through their digestive systems.

The most important effects of earthworm activity are the: (1) vertical and horizontal mixing of soil and (2) increased aeration and drainage resulting from burrowing. A widely held opinion is that earthworms increase soil fertility, but little evidence supports this claim. These macroorganisms tend to be restricted by low soil reaction, hence they are more abundant in soils with higher amounts of available calcium. Therefore, they can be considered as an indication, rather than a cause, of

---

[3] In the process of penetrating the plant root, nematodes open entrances into the plant for less efficient parasites.

soil fertility. In the process of digesting organic matter, humification takes place in the worm's stomach before the material is excreted. Altered in size and transported to lower soil levels, the ingested material is reduced to colloids and serves as an ion absorber in soil cation-exchange processes.

Arthropods. The number of arthropods in one-half hectare of well manured land has been estimated at near 8,000,000. Although individual insects are very small and provide little organic material, the short lives, rapid reproduction rate, and vast numbers make them a significant soil element. Not only do they constitute a part of the organic soil fraction, but they also can materially affect the mineral content of varied levels of soil through mixing. Thorp (1967, 196) gives a succinct description of soil alteration by the leaf-cutting ants of the tropics:

> The leaf-cutting ants march for long distances, cut out fragments of leaves and stems of plants that are to be used directly or indirectly for food, carry them back home, and store them in underground chambers. Here they are used to produce fungus that is in turn used for food. In this way organic matter, both in the forests of Central America and in the brushlands of northern Mexico and Texas, is incorporated in the soil and converted to humus through the activities of fungi (planted by the ants) and by the deposition of fecal matter by the ants themselves. In a private communication, R. L. Pendleton reports that mounds of these leaf-cutter ants in Central America frequently are as much as 15 feet [approximately 5 meters] across and 3 feet high [about 1 meter]. I have seen smaller ones in the brushy lands of south Texas. The mere tunneling operations of these ants have the effect of moving mineral material from one horizon of the soil to another, and after the ant hills are abandoned the chambers provide channels for rapid water penetration to the deeper subsoil horizon.

Another example of the arthropod's impact on the soil is demonstrated by the role of termites in the tropics. Termite mounds have been known to average from 2 to 3 meters in height and may be up to 17 meters in diameter, with as many as 75 mounds per hectare. Such construction is estimated to transport more than 2,000,000 kilograms of soil per hectare.

Other arthropods include the *mites* (microscopic to barely visible in size) and the *springtails*. Mites may reach populations of several billion per hectare. Some types feed on other small soil fauna, but most feast on organic matter—as do the springtails—and contribute significantly to the breakdown of organic residues and fungi.

Other soil fauna. Numerous larger animals also find soil a suitable habitat. Mollusks (snails and slugs), crustaceans (crabs and crawfish), reptiles (snakes and lizards), and mammals (gophers, ground squirrels, woodchucks, and other small vertebrates) are all effective in the total process of granulating, pulverizing, and transferring large amounts of soil and incorporating organic matter. Furthermore, by burrowing, they normally contribute to an increase in soil aeration and drainage.

A SUMMARY OF SOIL INHABITANTS

The life cycle of soil organisms may be said to interlock with that of higher plants, each requiring the other for survival. Habitat factors are extremely important to microflora and fauna. For development and growth, they must have sources of energy input, as well as many of the elements essential to higher plants. Energy is obtained either through the oxidation of inorganic substances (autotrophic bacteria) or from organic materials (heterotrophic bacteria, fungi, actinomycetes, etc.). Organic remains of higher plants are primary suppliers of the basic needs of most soil life forms. Decomposition of organic matter releases nutrients previously held in tissues, making them once again available for further plant growth. This prevents continued accumulation of vegetable debris. During the process, certain microbes, for example, bacteria, are able to fix free nitrogen. This capability should not be underestimated, because in its elemental form, nitrogen—necessary for good plant productivity—is unavailable.

Soil organisms also have other requirements in addition to energy and nutrient supplies. Some of the important ones that affect soil populations are aeration, moisture availability, pH (the degree of acidity or alkalinity), and temperature.

Microorganisms also influence the stability of soil aggregates. During intermediate stages of decomposition, microbial gums and slimy vegetative products cause soil particles to stick together and increase their resistance to breaking apart. Not all activities of the soil inhabitants, however, are beneficial. Crops may be destroyed by plant root infestations or burrowing animals, and soil organisms, if they are not limited, compete detrimentally with higher plants for available nutrients and oxygen.

SOIL ORGANIC MATTER

Organic matter refers to all living and dead matter within and upon the soil. The organic matter consisting of non-decomposed leaves, twigs, or stalks lying upon the surface is known as litter; plant and animal residues in the process of decomposition are called humus.

Soil organic matter content is determined by a combination of several factors: climate and vegetation, parent material, topography, and time.

1. *Climate and vegetation.* The distribution of world vegetation types is closely associated with specific patterns of temperature and precipitation. It is common knowledge, for example, that the sparse vegetation of desert regions is intimately tied to meager precipitation. In transitional climates, between desert and forested environments, a precipitation gradient exists— increasing in the direction of more humid, forested areas. Native vegetation of the transitional region is dominantly of the grass family. The height and completeness of surface cover and the luxuriousness of the grasses increase in the direction of increasing precipitation (see Figure 2.3). Similarly, as the grass cover and its associated root complex become more abundant, the organic content of soils increases. In grasslands of the United States this variation ranges from 200 metric tons of organic matter per hectare— within the top 100 centimeters of the soil—along the arid climatic boundry, to 400 metric tons per hectare within the humid, tall-grass prairie region (Foth and Turk, 1972). When precipitation exceeds environmental moisture requirements, a forest vegetation cover is usually present.[4] Organic content of forest soils is substantially less than in grasslands. The reason lies largely in the root structure of plants. Root systems of grass are concentrated near the surface and diminish gradually with depth, providing greatest organic content near to the surface and gradually decreasing downward. Roots of trees, on the other hand, are deeply distributed, yet trees supply most of their organic debris as leaf fall, concentrating the major dead plant material on the surface rather than incorporating it within the soil.

Another climatic influence determining accumulation of organic matter relates directly to temperature. In cold climates where soils are frozen for several months of the year, the rate of decomposition is extremely slow and organic matter accumulates. Consistently warm and moist climates favor development of active microbial populations that decompose material throughout the year, tending toward low soil organic content (Figure 2.4).

2. *Parent material.* Inorganic materials also influence the amount and type of vegetation growing on a particular site. Minerals supplying abundant nutrients, in particular calcium and phosphorous, foster more vigorous vegetative growth than that which occurs on sites deficient in

---

[4] An area's moisture requirement is generally considered a function of the energy available for evaporation and transpiration processes. This "moisture demand" is called *potential evapotranspiration* and is defined as the total amount of moisture which could be evaporated and/or transpired under conditions of optimal soil moisture.

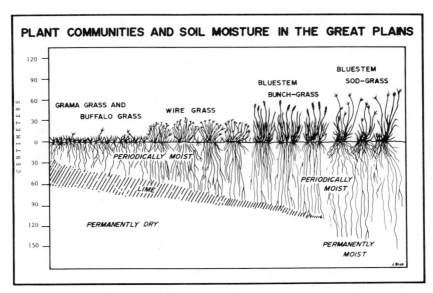

PLANT COMMUNITIES AND SOIL MOISTURE IN THE GREAT PLAINS

Figure 2.3 Relationship of plant communities to the depth of soil moisture penetration in the Great Plains of North America.

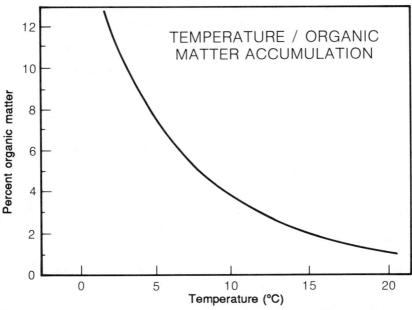

TEMPERATURE / ORGANIC MATTER ACCUMULATION

Figure 2.4 Organic matter accumulation bears an inverse relationship to mean annual temperature. Organic matter decomposition rates, however, increase with increased temperature.

plant foods. Because profuseness of vegetation varies, potential for the accumulation of organic matter in soil varies also.

3. *Topography.* The effect of surface relief on organic matter accumulation is primarily related to drainage conditions. Depressions toward which moisture drains have greater organic accumulations than adjacent, well-drained slopes. Factors contributing to this difference include (1) greater availability of moisture for plant growth, (2) downslope surface wash of fine-sized organic matter, (3) downslope migration of colloid-sized organic material in suspension within the soil, and (4) decreased oxidation of plant remains in depressions due to standing water dominating soil pore spaces for greater durations of time than on adjacent slopes.

4. *Time.* Vegetative cover and accumulation of soil organic matter are not instantaneous occurrences. Beginning with a bare surface and lacking nitrogen, plant growth is slow and organic matter inclusions are minimal. Only through time and weathering do nutrients become available to enhance vegetative growth and increase organic accumulations.

HUMUS

Humus is dark in color (usually brown or dark brown) and is a rather stable fraction of soil organic matter that remains after the major portion of plant and animal residue has decomposed. As mentioned earlier, processes involved in organic decomposition and that lend to formation of humus are called *humification.* An interesting aspect of humus development is that organic residues are selectively broken down by soil organisms. Thus, the composition of the organic matter is greatly changed from its original form as decomposition takes place. Plant tissue contains cellulose (20 to 50 percent); hemicellulose (10 to 30 percent); lignin (19 to 30 percent); protein (1 to 15 percent); and fats, waxes, and other substances (1 to 8 percent). In humus formation the celluloses and hemicelluloses are reduced rapidly. Lignin, being more resistant to destruction, is chemically altered but remains in the soil as a dominant residual. Proteins are also retained in significant quantities, but for different reasons. It is thought that protein molecules are either (1) absorbed on the surface of clay minerals and become resistant to decomposition or (2) are trapped by clay particles, rendering them inactive. The result of such preferential decomposition is a soil organic content with percentage remains as follows: cellulose (2 to 10 percent); hemicellulose (0 to 2 percent); lignin (35 to 50 percent); protein (28 to 35 percent); and fats, waxes, and others (1 to 8 percent).

Humus produces significant changes in soil by increasing the following:

1. Soil's ability to hold exchangeable ions. Humus of colloid size (less than 1 micron) has a high cation-exchange capacity—approximately twice that of clay. Cations such as Ca, Mg, and K are thus altered from leaching and maintained in a form available for microorganisms and higher plants.

2. Soil's water-holding capacity. Humus has the ability to absorb large quantities of water, shrinking and swelling with moisture changes.

3. Soil's carbon and nitrogen content. The carbon content of humus theoretically averages at 56.24 percent, and nitrogen content, 5.6 percent. The ratio of these two is 10.04:1, which approximates very closely the ratio found in numerous soils. Determination of the *carbon-nitrogen* (C–N) *ratio*, which is relatively constant for soils within a given climatic regime, provides an index of the breakdown rate of organic matter by microorganisms. Microbial enzymatic processes convert organic carbon to gaseous carbon dioxide to obtain energy, yet these organisms are inefficient in incorporating the carbon into their cell structure. As decomposition proceeds, considerable carbon is released into the atmosphere, whereas practically all of the nitrogen is incorporated into new protein molecules, resulting in a narrowing of the carbon-nitrogen ratio.

Within soil the practical significance of the C–N ratio relates to competition for nitrogen between higher plants and the microbes. When organic residues with a wide C–N ratio are soil incorporated, microbial activity is stimulated, soon reaches its peak, monopolizes available nitrogen, and leaves plants nitrogen deficient.

4. Soil's *tilth. Tilth* refers to the physical conditions of soil as related to ease of tillage, fitness as a seedbed, and resistance to emerging seedlings and root penetration.

5. Soil's structural development. Active microbial activity associated with humus production results in mycelial growth and microbial gums capable of binding soil separates together.

ORGANIC MATERIAL AND SOIL COLOR

In Chapter 1, soil color was considered solely in terms of the impact of chemical weathering on inorganic materials. Organic matter, however, substantially contributes to the color of a soil and can be sufficiently dominant to mask inorganic coloration. A high humus content in pore spaces and as coatings on mineral particles normally gives soil a dark or black color. Prairie soils, where luxurient growths of grasses occur, are usually dark colored. Low organic content, on the other hand, leads to lighter inorganic coloration, typical of desert soils. A more comprehensive discussion of soil color is provided in Appendix III.

Soil Porosity, Moisture, and Atmosphere

3

An ideal loam soil is comprised of approximately 45 percent mineral matter. The remainder consists of about 5 percent organic matter and 50 percent pore space, about 25 percent of which is occupied by air or water. Air and water are thus variable elements in the soil complex, the quantity of one being inversely related to the other. When soil is saturated with water after a heavy rain, air content may be near or at zero. If drought conditions persist for several months, soil atmosphere increases as moisture reserves diminish.

PORE SPACE

Water and air are found in the interstitial spaces between soil particles. The supply of water, its rate of movement, and the availability of oxygen are all determined largely by the amount and size of soil pores. To establish a given soil's pore space, pedologists must consider both the *bulk density* and *particle density* of soil material.

BULK DENSITY

The *weight* of an "oven-dry" soil sample divided by its volume constitutes bulk density, or

$$\text{Bulk density (BD)} = \frac{\text{Weight (grams)}}{\text{Volume (cm}^3)}$$

The greater the bulk density, the less the pore space. For example, solid rock material with no pore space has greater weight per volume, hence greater bulk density, than the same volume of shattered rock material of similar composition (Figure 3.1). The reason for the difference is an inability of irregular rock fragments to fit together perfectly, thus creating open space between fragments and decreasing total weight of the mass per unit volume. (Figure 3.1)

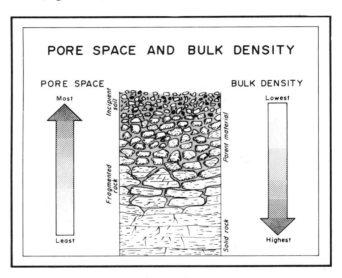

Figure 3.1 Pore space and bulk density vary inversely. As inorganic soil components decrease in size, the total pore space normally increases and bulk density decreases.

PARTICLE DENSITY

Unlike bulk density, which includes solids and pore spaces, particle density is weight per unit volume of soil solids or weight of solids only, as if the soil particles were compressed into a volume wherein all pore spaces have been squeezed out.

$$\text{Particle density (PD)} = \frac{\text{Weight of soil solids (grams)}}{\text{Volume of soil solids (cm}^3)}$$

Note the difference between bulk density and particle density in the examples provided in Figure 3.2.

BULK DENSITY VERSUS PARTICLE DENSITY

Calculation of Bulk Density

Volume of soil sample = 1 cm.^{3}

Weight of soil sample = 1.5 gms.

Bulk Density = $\frac{1.5 \text{ gms.}}{1.0 \text{ cm}^3}$ = 1.5 gms./cm.^{3}

Calculation of Particle Density

Volume of compressed soil sample
= 0.55 cm.^{3}

Weight of compressed soil sample
= 1.5 gms.

Particle Density = $\frac{1.5 \text{ gms.}}{.55 \text{ cm}^3}$ = 2.7 gms./cm.^{3}

Solids and
Pore Space

1.5 gms.

Volume = 1 cc.

Compressed solids
(No pore space)

1.5 gms

Volume = 0.55 cc.

Figure 3.2 Examples of the calculation of bulk density and particle density.

PORE SPACE SIZE AND AMOUNT

Knowing bulk and particle densities, it is possible to determine percentage of pore space in mineral soils, that is, the amount of soil occupied by either air or water. Obviously, the total volume occupied by a given soil is a sum of pore space and solid particles. If the percentage value of either soil portion is known, the other is easily calculated. The following equations are useful for obtaining a pore space value:

$$\text{Soil solids } (\%) = \frac{\text{BD}}{\text{PD}} \times 100$$

$$\text{Pore space } (\%) = 100 - \% \text{ Soil solids}$$

Size of pore space varies with texture. Large soil separates have large pores, whereas small particles, such as clay, have micropores. Conversely, total pore space is greater and bulk density generally lower in fine-textured soil material. This apparent paradox is easily explained. Pore space size is directly proportional to particle dimensions. The smallest mineral components, although having minute interstitial spaces, have greater total pore space because a greater number of particles are in contact with one another, per unit volume. As fine-sized materials become more dominant within soil, pore space size decreases further and total pore space continues to increase (Figure 3.1). An exception occurs in subsoils with many different particle sizes. Interstitial spaces can become clogged with smaller or colloid-sized materials and/or translocated clays. When soil is compacted by machinery during cultivation, smaller soil particles fill pore spaces between larger particles.

Pore space size affects moisture infiltration and storage and soil aeration. Large voids—*macropores*—allow rapid infiltration of rainfall or irrigation water into the soil. This water, however, will also rapidly drain out of large pores. Moisture storage thus tends to be low in coarse-textured soils. This explains the presence of drought-resistant plant species (*xerophytes*) on sandy soils within humid climatic regions. Since a plant's water needs are largely derived from stored soil moisture, meager reserves result in occupancy by environmentally adapted flora capable of surviving in dry sites where other plants suffer stress and wilt.

Small pores—*micropores*—typical of clay soils restrict water infiltration into soil. Water moves slowly through the minute void spaces. Drainage is also slow, indicating a higher soil water-holding capacity.

Since water drains from large pores readily, macropores are usually filled with air and referred to as *aeration* pores. Micropores are normally so small that the capillary films of moisture surrounding each soil particle saturate adjacent void spaces with water. Hence, micropores are commonly called *capillary pores*. The size of pores and total amount of pore space are important soil characteristics.

SOIL MOISTURE

Rainfall delivers atmospheric moisture that eventually rests upon land surfaces, within soil (soil water), or beneath the earth's surface in the form of ground water. To become soil water, precipitation must first infiltrate the soil crust and percolate downward via gravitation and capillary forces.

Infiltration is a process whereby water enters soil via surface void spaces and subsequently percolates downward. Each soil has a unique ability to infiltrate or absorb specific amounts of water throughout given time periods. This absorption potential is the soil's *infiltration capacity*. Clearly, rain enters the soil at capacity rates only during or immediately following periods when precipitation equals or exceeds the infiltration capacity.

Numerous factors operate simultaneously to affect the infiltation capacity from one location to another or from one season to another. They include the following:

1. *Soil moisture.* If soil is dry at the initiation of a rainfall event, wetting its top layer generates a strong capillary potential immediately below the soil surface, thus supplementing gravitational force and

increasing infiltration. On the other hand, when colloids—organic and inorganic—and clay particles absorb water, they may swell and reduce the infiltration rate.

2. *Compaction due to rain.* Sandy soils are little affected by rain compaction. Fine-textured soils, such as clays, are subject to raindrop impact, which compresses soil particles and reduces pore space and infiltration.

3. *Inwash of fine materials.* Colloids are held in suspension by rainwater on soil surfaces. As infiltration begins, the soil acts as a filter, collecting the colloids in pore spaces. As accumulation continues, interstitial spaces often become clogged and infiltration is reduced.

4. *Topography.* Slope gradient, length, and shape all determine how rapidly water runs off the surface and, consequently, how long a given volume of water will remain on a soil surface to infiltrate the soil.

Other factors are also important. Among them are: (1) viscosity (gluey consistency) of the infiltrating water, (2) proportion of air trapped in subsurface pore spaces, (3) degree of compaction and inwash reduction by vegetative cover, and (4) amount of soil compaction due to animals. Seasonal variation of infiltration rates may occur due to changes in land use, characteristics of vegetation, cultivation practices, and temperature. These factors illustrate that infiltration capacity can significantly differ within short distances and through time.

Once water infiltrates the soil surface, it moves into the main body of the soil, enveloping in a film of water each soil particle it contacts, filling micropores. If available in sufficient quantity, infiltrating water occupies macropores as well.

Soil moisture occurs in vapor form or as adhesion, cohesion, and gravitational water. Understanding these moisture forms and the forces retaining them in soil provides knowledge of water's subsurface movement and availability to plants.

Water of adhesion is a moisture film having immediate contact with soil particles. Several molecules thick, it is rigidly held by adhesive forces (an electrical attraction existing between water molecules and soil separates) and under normal conditions is found in all soils, including those of the desert. Most pedologists consider water of adhesion to be in a crystalline state, exhibiting insignificant movement, or none at all, and having low available energy. To remove this moisture from soil, a sample is oven-dried under high temperatures.[1] Its quantity is determined by

---

[1] Soil samples are oven-dried at 105°C for a period of 24 to 48 hours.

establishing the soil sample's air-dry weight relative to its post oven-dry weight.

Water of cohesion, also known as *capillary water,* both encircles each soil particle (and its associated adhesion water film) with a liquid water envelope and also occupies micropore space. This water can have an approximate thickness of fifteen molecular layers. Compared with adhesion water, it has more energy, greater internal mobility, and increased soil movement potential, and is a source of plant water needs. In fact, plants obtain their primary water and nutrient needs from water of cohesion, even though only about 66 percent of the outer molecular layer of this water film is available for their use. Clearly, the force retaining this moisture exceeds that of gravity. Otherwise, the water would drain out under the influence of gravity. The level of energy with which cohesion water is held in soil is referred to as *matric potential.* (*Matric* can be considered equivalent to capillary. The origin of the term is from the word *matrix,* meaning the mineral separates that constitute the soil body.) When micropores are saturated and soil particles are surrounded by their maximum amount of cohesion water, the matric potential, that is, the capillary attraction force, is zero. As water is removed and the capillary film thins, soil moisture is held with increasing tension and requires greater energy availability to be used for plant root absorption and evaporation processes.

Matric potential is commonly expressed as bars (or atmospheres), a term familiar to those who have studied meteorology. In weather terminology, a bar approximates the *standard atmospheric pressure,* which equals 14.7 pounds per square inch, or a barometer reading of 760 millimeters of mercury. In water bodies, such as lakes, a pressure equivalent to 1 bar exists at a depth of 1,020 centimeters due to the weight of water as influenced by gravity. This hydraulic pressure increases downward and decreases upward at the rate of 0.00102 bars per centimeter. At the surface, the lake's hydraulic pressure is zero; otherwise it is always positive.

In soil, matric potential is measured in negative atmospheres ($-$bars), except during saturation, when the pull of capillary force is, zero atmospheres (i.e., similar to the hydraulic pressure of a free water surface). When capillary films are partially reduced, matric potential is reduced from zero and can only be negative, unlike the situation with an open body of water. From saturation to air-dry status, matric potential can range from 0 to $-1,000$ bars.

Plants obtain most of their water needs from the cohesion layer; yet, as previously discussed, not all of this moisture can be utilized. When matric potential approaches -15 bars, the rate of moisture uptake by plant

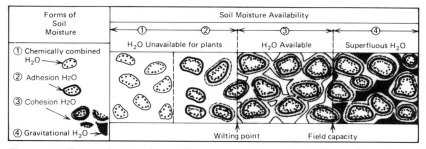

Forms of Soil Moisture	Soil Moisture Availability			
	←——①——→|←——②——→| |←——③——→| |←——④——→|
① Chemically combined H_2O →⬥	H_2O Unavailable for plants		H_2O Available	Superfluous H_2O
② Adhesion H_2O →⬥				
③ Cohesion H_2O →⬥				
④ Gravitational H_2O →⬥		Wilting point	Field capacity	

Figure 3.3 Forms of soil moisture. The amount of water available to plants is equal to the field capacity minus the wilting point.

roots falls short of water lost via transpiration, and *wilting* normally occurs.[2] Hence, the water content of a soil with matric potential of -15 bars is an estimate of that soil's *permanent wilting point*.

GRAVITATIONAL OR FREE WATER

Following a prolonged rain or period of irrigation, the soil may have all interstitial spaces occupied by water; it is then considered *saturated.* Much soil moisture at this stage, however, is occupying macropores, which, having very little capillary potential, will rapidly drain once the external water supply ceases. Such water, held only temporarily within the soil, is referred to as *gravitational* or *free water*. Because it is present for only a very short period before draining from the soil, its use by plants is quite limited. After gravitational drainage occurs, soil is said to be at *field capacity.*

AVAILABLE WATER

As noted previously, the smaller the particle size, the greater the number of particles and total pore space of the soil. With increased numbers of particles and surface area and increased moisture-holding capacity of the soil, there are more particles to hold films of both adhesion and cohesion water, and fewer macropores from which water will drain. Hence, soil water storage is directly dependent upon particle size, as well as structure and organic content.

Available water is considered to be the difference between a soil's *field capacity* and the *wilting point* of plants (Figure 3.3). When fine particles

---

[2] *Wilting* occurs to plants when their water intake is insufficient to replace that lost by transpiration. There is a loss of turgidity in plant tissue associated with deflation of plant cells.

dominate soil, the field capacity increases and the amount of water of adhesion also increases. The proportion of available water for plants to the total water held by the soil, however, decreases (Figure 3.4).

The volume of available water in the *rhizosphere* (root zone of soil) at any moment is largely dependent on regional climate patterns. Except for artificially adding water through irrigation, the primary source of soil moisture comes from naturally occurring precipitation, which infiltrates and replenishes the soil reservoir. Each climatic zone, in turn, imposes a demand for moisture upon the soil. This water requirement is induced by solar energy, which is capable either of converting free soil water in its liquid state to a gaseous form (evaporation) or converting water existing within pore spaces of plants to a gaseous state (transpiration). The total capacity for such atmospheric moisture transformations is called *potential evapotranspiration* (PE), in other words, the amount of water that could be evaporated and transpired under conditions of optimal soil moisture and specific availability of atmospheric energy.

Varied precipitation and thermal zones occur over the earth's surface, resulting in regional differences of moisture supply (precipitation) and demand (potential evapotranspiration). Thus while vegetation in one region can survive the short periods between precipitation by extracting soil moisture without experiencing moisture stress, other areas are characterized by seasonal soil water depletions sufficiently severe to require physiognomic adaptations by plants.

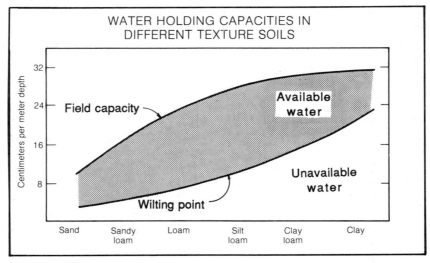

Figure 3.4 General relationship between soil moisture availability and soil texture.

A simple graphic method for describing the moisture supply and demand situation is shown in Figure 3.5 for two locations in North America. Brevard, North Carolina, has a low winter moisture requirement (less than 12 millimeters per month) due to low temperatures and short daily periods of illumination. This need increases rapidly in the spring as solar radiation becomes more intense and the period of illumination lengthens, reaching a peak of over 125 millimeters in July. It then drops rapidly in autumn. In each month, precipitation exceeds the moisture

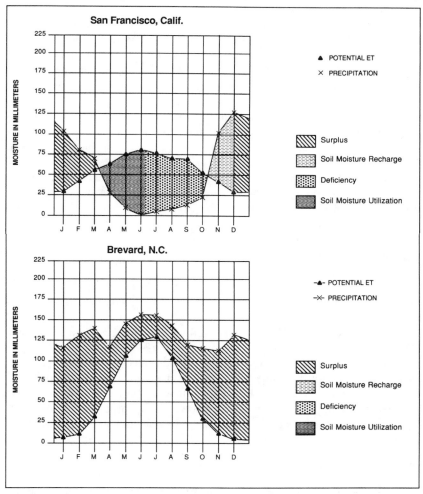

Figure 3.5 Water budgets of Brevard, North Carolina, and San Francisco, California.

demand. As soil moisture is utilized, it is replenished, and excess rainfall either drains through the soil or is lost as surface runoff.

The moisture requirement of San Francisco, California, is higher in winter and lower in summer than that of Brevard. Although San Francisco's summer moisture demands are less, the ability of the atmosphere to supply rainfall is exceedingly low. Hence, transpiration and evaporation heavily tax the soil reservoir until the moisture supply is nearly exhausted and a water deficiency exists.[3]

Interrelationships between precipitation and potential evapotranspiration are vital aspects of regional *pedogenic* (soil formation) processes. Although this topic is explored in detail in later chapters, a few brief examples of the far-ranging effects of moisture supply and demand are appropriate here: (1) Where precipitation exceeds potential evapotranspiration, the chemical alteration of parent material is enhanced by frequent soil leaching; (2) the character of vegetation, hence the organic content of soil, is subject to the moisture supply and thermal regime; and (3) dehydration of soils during periods of moisture deficiency can lead to the precipitation and accumulation of minerals previously maintained in soil solutions.

Soil depth, texture, organic matter content, and structure—in combination—dictate the amount of water capable of being stored in a given soil. These soil components can also influence rate of soil moisture release to the atmosphere and, more interestingly, they themselves are products of the moisture-energy regime in which they play a role. A continuing short-term (weekly or monthly) computation of moisture supply and demand relationships or *water budget* (precipitation—potential evapotranspiration—soil moisture storage balance) of a site provides a current measure of moisture available within the rhizosphere. Such datum is beneficial; for instance, informing an agriculturist when to replenish soil moisture through irrigation before soil water depletion has seriously harmed crops.

Figures 3.6 and 3.7 illustrate available moisture and its seasonal variation by area in the United States. Both maps incorporate the potential evapotranspiration measure. Remember that it is impossible to know whether a climate is moist or dry by knowing only precipitation amounts. It is necessary to compare precipitation with potential evapotranspiration.

---

[3] The rate of moisture release from the soil is dependent upon the quantity of available water in the soil reservoir. At field capacity, release approximates demand. As the soil dries out, moisture release becomes a function of the percentage of remaining soil moisture and energy demands; that is, at 20 percent field capacity, release may approximate only 20 percent of demand.

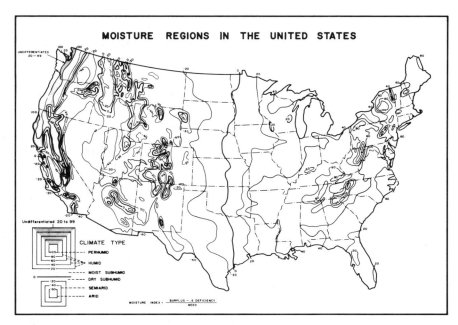

Figure 3.6 Moisture regions in the United States (after C. W. Thornthwaite).

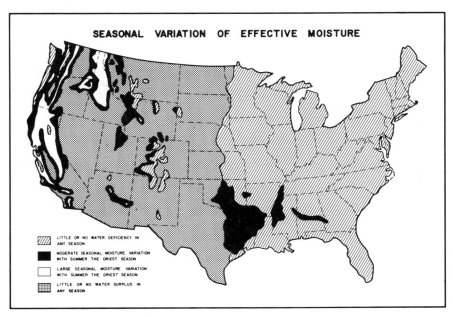

Figure 3.7 Seasonal variation in effective moisture (after C. W. Thornthwaite).

Likewise, such moisture-energy interrelationships must also be considered in order to understand the effect of climatic variables in soil formation and upon moisture available to plants.

SOIL ATMOSPHERE

A *soil atmosphere* of unique gaseous composition exists in soil pore spaces that are not occupied by water. Unlike free atmosphere, soil atmosphere is rich in carbon dioxide (10 to 100 percent greater than above the surface) and has lower quantities of oxygen. The reason is that plant roots and organisms living in soil remove oxygen from, and respire carbon dioxide into, the interstitial spaces. Most crops cannot grow if carbon dioxide content in the root zone is too high or oxygen is too low. A constant exchange between the soil and free atmosphere must take place. The free atmosphere supplies oxygen, and the soil diffuses carbon dioxide into the free atmosophere. For soils to be well aerated, sufficient open interstitial spaces must be available to permit easy movement of essential gases in and out of them. Saturated soils obviously have little space available for air.

Effect of Site and Time on Soil Characteristics

4

The major components of soil—inorganic material, organic material, water, and air—have been individually considered. It is obvious that dominance of one component over another can produce varied soils. Yet the question remains, why do soils occurring on similar parent material have different characteristics in one location from those adjacent to or distant from them? One obvious factor is biotic. In Chapter 2 it was shown that organic content and coloration of soil were strongly associated with type of vegetative cover. Yet, there are other aspects of *pedogenesis* (soil formation) that are also significant in producing variation. The most important are topography, climate, and length of time soil development has been taking place.

TOPOGRAPHIC POSITION

Topographic position is an important variable in determining the nature of a given soil. Landscapes in mountainous regions exist in a number of climatic and vegetative zones due to altitudinal variations in both temperature and moisture. As a consequence, exposed parent materials of similar composition within relatively short distances of one another may be influenced by different sets of climatically related, soil-forming processes. Furthermore, local relief and drainage characteristics affect soil properties even within a uniform climatic regime.

ALTITUDE DIFFERENCES

In mountainous regions, steep slopes can speed removal of loose surface materials by enhancing the erosive power of gravity and running water, thus not permitting deep soil development. The mountain soils of Arizona have been described by Buol (1966, 3):

> These soils result largely from the interaction of topography and time. The steepness of the slopes allows for the rapid removal of friable soil material by water and gravity. Therefore, the soil surface is constantly lowered at such a rapid rate that many of the soil-forming processes controlled by climate and organisms do not have time to impart a significant influence on the soil development. Instead, most of the soil characteristics—namely, texture and chemical composition—are closely related to those of the parent rock.
>
> The influence of climate and vegetation is observed mainly in the color and organic content of the surface layer of the soil. The cooler and wetter areas, which usually support more vegetation, have a higher organic matter content, and thus a darker color in the surface layer, than where the climate is hotter and drier. Organic matter is also oxidized at a slower rate in a cold climate than in a warm one.
>
> Being shallow over rock material, these soils store little water to be used by growing plants; consequently, unless rains are frequent, vegetation is sparse.

Mountainous regions exist in a variety of climatic regimes that are the direct result of decreasing temperature and increasing orographic (mountain-induced) precipitation that increases with the rise in elevation above sea level. Figure 4.1 illustrates this relationship, as it exists, in the Catalina Mountains near Tucson, Arizona. Associated with vertical, climatic zonation is a variety of floral habitats, ranging from the hot subtropical desert's scrub to the humid and cool high-elevation zone's fir forests, as graphically described in Figure 4.2 for the central Sierra Nevada.

The result of varied vegetation and climate interacting with parent material is a vertical succession of soil types with differing features. An examination of soils in the mountains of the southwestern United States reveals the following:

1. Organic content of the soils increases with elevation.

2. Total porosity and water-holding capacity—associated with organic matter—increase with elevation.

3. Values of pH decrease with increased elevation; desert soils are alkaline and calcareous (containing calcium or a calcium compound) layers, and high mountain soils are more acidic.

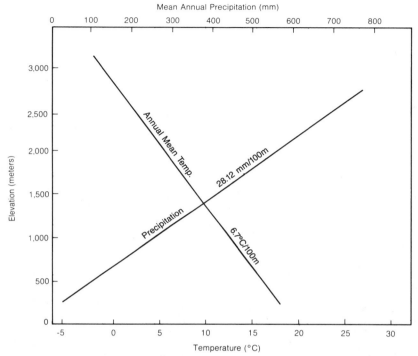

Figure 4.1 Relationship of elevation with precipitation and mean annual temperature in the Catalina Mountains of southern Arizona.

4. C–N ratio increases with elevation.

Local Relief

Soils with identical parent material, existing within a uniform climatic region and affected by parallel soil-forming factors, may still have varied features due to local relief and/or internal drainage attributes. The name for such a group of related, yet different, soils is *catena,* meaning "chain."[1] Figure 4.3 demonstrates the catena concept for a valley-slope profile in the Machkund Basin of India. Soil boundaries are marked and each soil classified by number. Table 4.1 contains descriptions of each soil, from which the following catena observations have been made (Rajan and Biswas, 1971, 80):

---

[1] A *toposequence* is also a small soil association occurring in the same climatic zone, under similar vegetation, and exhibiting differences related to topography. However, these associated soils do not have to possess a common parent material.

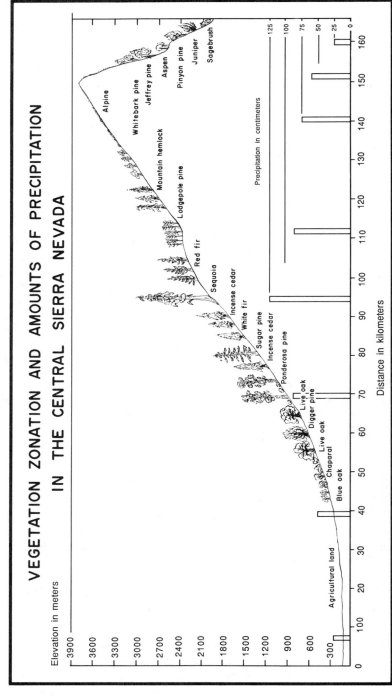

Figure 4.2 Relationship between elevation, precipitation, and vegetation.

56

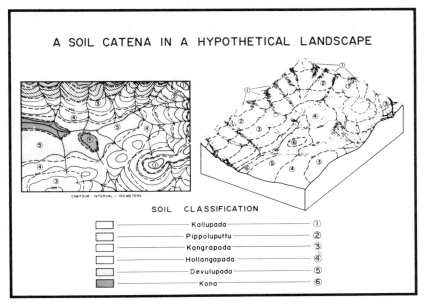

A SOIL CATENA IN A HYPOTHETICAL LANDSCAPE

CONTOUR INTERVAL – 100 METERS

SOIL CLASSIFICATION

☐	———————— Kallupada ————————	①
☐	———————— Pippoluputtu ————————	②
☐	———————— Kangrapada ————————	③
☐	———————— Hollangapada ————————	④
☐	———————— Devulupada ————————	⑤
▨	———————— Kona ————————	⑥

Figure 4.3 A soil catena in a hypothetical landscape.

The soils of the hilltops and the slope of hillsides, namely, soils of the Kallupada series, are shallow, occasionally attaining moderate depth. As the slope eases up the foothill areas, the soils attain greater depth. The soils of the valley and depressions are very deep. The data of mechanical and chemical composition of the soils . . . show that there has been an increase in the content of clay in the soils from hilltop to valley.

From the foregoing discussion, it would be apparent that there has been lateral translocation of finer materials as also of soluble constituents on account of lateral movement of water, which flows down as runoff. The above situation is suggestive of lack of deep percolation through depth of hilltops and slopes, giving rise to shallow soil. Further accelerated erosion due to slope conditions depletes the soil materials from convex and steep-slope situations, giving rise to shallow soils. Deep percolation of water in the other areas accentuating weathering and receipt of soil material from the upland and through the drainage courses have given rise to deep soils in valleys and depressions.

The cation exchange capacity of the soils also varies, being lower in respect of soils of the hills and hillslopes and higher in respect of the soils in the low-lying situation. The high content of dibasic constituents in the low-lying soils coupled with high moisture regime appears to have helped resilification of the secondary clay minerals enriching the soils in contents of 2:1 lattice minerals of high-exchange capacity. The derived values of cation exchange capacity definitely show a transition from hills to low-lying areas.

Table 4.1   Physical and Physio-Chemical Properties of Horizon Soil Samples of the Recognized Soil Series in the Machkund Basin

No.	Soil series and depth (cm)	Mechanical constituents (expressed as percentage on air-dry basis)			Physio-chemical properties				
		Sand	Silt	Clay	Total c.e.c.[a] m.e.[b] 100 g	T.E.B.[c] m.e. 100 g	Base saturation (%)	pH	B.E.C.[d] m.e.% calculated on clay
1.	Kallupada								
	0-10	71.4	9.0	19.9	6.7	3.2	47.5	5.4	33.8
	10-35	54.3	18.5	27.1	8.8	4.8	54.5	5.6	32.4
	35-45	45.8	21.3	32.8	9.2	4.8	52.1	5.6	28.1
2.	Pippoluputtu								
	0-12.5	73.9	10.3	15.8	7.2	3.8	52.7	5.4	45.4
	12.5-40	61.6	13.7	24.7	8.4	4.0	50.9	5.5	34.0
	40-75	48.4	14.2	37.4	10.8	5.5	51.1	5.7	29.0
3.	Kangrapada								
	8-15	55.8	21.4	22.7	6.8	3.5	51.8	5.6	30.0
	15-42	49.9	22.6	27.5	8.9	4.5	50.9	5.6	32.5
	42-90	44.5	22.9	32.6	10.6	5.5	51.7	5.7	32.7
4.	Hollangapada								
	0-15	56.7	21.1	22.1	8.4	4.3	51.4	5.4	38.0
	15-57.5	58.7	17.2	24.0	9.3	5.1	54.4	5.4	38.7
	57.5-150	58.9	15.6	25.4	10.8	5.7	52.7	5.6	42.4
5.	Devulupada								
	0-12.5	53.9	19.5	26.6	8.8	5.2	59.0	5.4	33.1
	12.5-45	49.8	20.8	29.4	13.2	7.9	60.1	5.4	44.9
	45-85	43.9	21.8	34.2	22.9	13.3	58.1	5.6	66.9
	85-150	35.6	23.6	40.7	26.3	15.7	59.9	5.6	64.3

Source: S. V. Govinda Rajan and N. R. Datta Biswas, "Development of Certain Soils in the Subtropical Humid Zone in Southeastern Parts of India, Genesis and Classification of Soils of Machkund Basin," *Soils and Tropical Weathering* (Paris: UNESCO, 1971), p. 83.

[a] c.e.c. = cation exchange capacity
[b] m.e. = milliequivalent
[c] T.E.B. = total exchangeable bases
[d] B.E.C. = base exchange capacity

THE SOIL PROFILE: A FACTOR OF TIME

Soil is a naturally occurring body experiencing continuous modification. Its properties are formed through the combined effects of *climate* and *biotic activity* (plants and animals) acting upon *parent material,* as conditioned by *topography* throughout periods of *time.*

When soil scientists analyze a given soil within a natural setting, they concern themselves with variation in both its physical and chemical traits with depth, as well as lateral extent. Our attention is first directed toward the *soil profile,* which is a description of a vertical, two-dimensional slice extending from the earth's surface through altered layers and into the nonsoil or parent material below. Figure 4.4 illustrates a succession of soil profiles, ranging from youthful (little or no development) to mature, that is, exhibiting differentiated properties with depth.

Soil description, as it varies with depth, provides information regarding plant root environments, nutrient and moisture availability, soil susceptibility to surface erosion and internal leaching, clay mineralogy, and other data upon which wise land-use planning may be based. Each *soil individual* is the synthesized product of all soil-forming factors operating *in situ* and has unique characteristics associated with its particular stage of development.[2]

Exposed crustal materials are subject to weathering processes that produce physically and chemically altered inorganic debris. Such material—given sufficient time and favorable atmospheric conditions—can have established upon and within it a community of plants and animals. The accumulation of their organic residue provides a new complex between the earth's geologic crust and the atmosphere—a body of material consisting of both organic and inorganic matter. This complex is conducive to support of an expanded biologic community. Microorganisms, including bacteria, protozoa, and a host of other flora and fauna, become increasingly more abundant, feeding on organic remains and releasing nutrients for further plant growth. Examination of soil at this stage of development reveals a surface layer that is darkened in color because of included organic matter in various states of decay overlying the parent material.

If soil-forming processes continue and rate of soil loss by erosion is less than the rate of rock weathering, soil will continue to deepen. As deepening takes place, soil composition is differentiated vertically. From the surface layer (topsoil) of mineral soil, under the influence of

---

[2] A *soil individual* is a soil distinct in characteristics from adjacent soil bodies or nonsoil materials. *In situ* (Latin) means in its original place.

percolating water, soluble minerals and suspended colloid-sized particles will have *eluviated* (removed). The colloids—clay, organic matter, and oxides of aluminum and iron—move only a few feet at most before they become lodged, creating an illuvial (accumulation) layer in the subsoil.

SOIL HORIZONS

Soil horizons are layers of soil or soil material lying approximately parallel to the land surface. They differ from adjacent genetically related layers in physical, chemical, and biological properties or characteristics, such as color, structure, texture, consistency, kinds and numbers of organisms, and degree of acidity or alkalinity. The number and variety of horizons indicate stage of soil evolution. Well-developed (mature) and undisturbed soils normally exhibit a sequence of horizons from the surface downward that are classified (in the pedologist's short-hand) as O, A, E, B, C, and R (Figures 4.4 and 4.5). These are known as master horizons and are individually described in Table 4.2.

In many soils, master horizons can be easily differentiated from one another and boundaries between them clearly identified. In other instances, changes in horizon traits are not obvious. When a gradual change occurs from one horizon to the next, the soil layer is considered a *transitional*

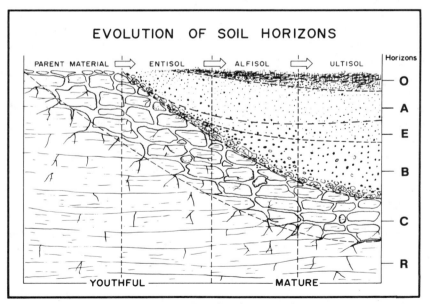

Figure 4.4 A sequence of soil profiles from youthful through mature.

Table 4.2 *Master Horizons*

Horizon designation	Description
O	Organic horizons of minerals soils. Horizons: (1) formed or forming in the upper part of mineral soils above the mineral part; (2) dominated by fresh or partly decomposed organic material; and (3) containing over 30 percent organic matter if the mineral fraction is greater than 50 percent clay, or over 20 percent organic matter if the mineral fraction has no clay. Intermediate clay content requires proportional organic matter content. The organic matter may be grass or forest debris lying on the surface of mineral soil, or it can be organic material that accumulated in poorly aerated water, as is necessary for the formation of *peat* or *muck* soils.
A	Mineral horizons having: (1) humified (decomposed) organic matter resulting from *in-site* plant decomposition or incorporation of O-horizon debris by the burrowing activities of soil animals, and (2) lost weathering products to lower horizons through eluviation processes.
E	Mineral horizons found either below an A or O horizon that have lost clay, iron, or aluminum with a resultant concentration of quartz or other resistant minerals of sand or silt size.
B	Horizons in which the dominant feature or features are one or more of the following: (1) an illuvial concentration of silicate clay, iron, aluminum, or humus, alone or in combination; (2) a residual concentration of sesquioxides or silicate clays, alone or mixed, that has formed by means other than solution and removal of carbonates or more soluble salts; (3) coatings of sesquioxides adequate to give conspicuously darker, stronger, or redder colors than overlying and underlying horizons in the same sequum[a] but without apparent illuviation of iron and not genetically related to B horizons that meet requirements of (1) or (2) in the same sequum; or (4) an alteration of material from its original condition in sequums lacking conditions defined in (1), (2), and (3) that obliterates original rock structure, that forms silicate clays, liberates oxides, or both, and that forms granular, blocky, or prismatic structure if textures are such that volume changes accompany changes in moisture.
C	A mineral horizon or layer, excluding bedrock that is either like or unlike the material from which the solum is presumed to have formed, relatively little affected by pedogenic processes, and lacking properties diagnostic of A or B but including materials modified by (1) weathering outside the zone of major biological activity; (2) reversible cementation, development of brittleness, development of high bulk density, and other properties characteristic of fragipans; (3) gleying; (4) accumulation of calcium or magnesium carbonate or more soluble salts; (5) cementation of accumulation of calcium or magnesium carbonate or more soluble salts; or (6) cementation by alkali-soluble siliceous material or by iron and silica.
R	Underlying consolidated bedrock, such as granite, sandstone, or limestone.

[a] *Sequum* is a sequence of an eluvial horizon and its subjacent B horizon, if present.

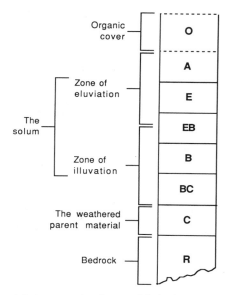

Figure 4.5 An example of a possible horizon sequence.

horizon and is designated by combined symbols, such as, AB, EB, or BA. In these instances, the transitional horizon AB has traits of both the A and B horizons, but looks more like A than B. Similarly, EB looks more like E than B and BA looks more like B than A. The kind and number of soil horizons formed are determined by the type and intensity of soil-forming processes that are operative within a specific environmental setting. Figure 4.5 illustrates some possible horizon subdivisions.

The pedologist's use of short-hand symbols also includes a consideration of unique soil traits that further refine horizon descriptions. Utilizing lower-case letters (a, b, etc.) as modifiers to master horizon symbols, for example, Bh or Bt, they identify significant soil attributes that are not represented by master horizon definitions. In the former examples, the lowercase letter *h* refers to the presence of humus, whereas *t* identifies a substantial accumulation of translocated silicate clay. A list, along with definitions, of each master horizon modifier is provided in Appendix II.

It is useful at this point to refer once again to Figure 4.4, which illustrates a sequence of soil profiles (differentiated by degree of evolution). Soil becomes reality when geologic materials are modified to the extent that they support higher forms of plant life. Initial modification of parent material results in soil that is only a slightly modified version of the environmentally exposed geology and is classified as an A horizon overlying C or R, that is, a sequence of A to C or A to R. The presence of both eluvial and illuvial layers—the soil's A, E, and B horizons—indicates advanced stages of soil evolution. If soil possesses only an A horizon resting upon partially altered parent material, it is said to be *young,* or *immature.* There has not been sufficient time in the given climate, along with related soil-forming processes, to produce horizon differentiation. A soil attains a *mature* stage of development with the formation of the eluvial E and illuvial B horizons (Figure 4.4). The A, E, and B horizons are collectively known as the *solum,* or *true* soil.

The concept of mature soil with horizon differentiation is extremely important to our understanding of global variation of regionally distributed soils. Parent material, topography, and time may all be viewed as passive factors in soil development, whereas climate and vegetation are active agents. A strong regional relationship exists between climate and vegetation, for example, humid regions and forests, semiarid areas and grasslands, and arid climates and deserts. Hence, it is expected that pedogenetic processes similarly occur on a macroscale over broad regions of the earth, producing mature soils that have similar traits throughout. This world pattern of mature soils and their classification is discussed in Chapters 5 through 12.

SOIL PEDON AND POLYPEDON

Profile was defined earlier as a two-dimensional vertical slice through the soil. Yet, soil is a three-dimensional entity with characteristics that either persist or vary along its *X, Y,* and *Z* axes. The smallest aggregate of particulate separates constituting a representative soil sample from some location within a profile has been previously defined as a *ped.* Combinations of peds constitute a *pedon* (Greek *pedon,* ground): a three-dimensional unit of soil with a lateral area large enough to permit the study of all horizon shapes and relations. Its area encompasses a minimum of 1 square meter (approximately 1 square yard). If soil horizons are laterally discontinuous or exhibit wavelike variation in thickness, however, the area needed to define a pedon may increase to as much as 10 square meters (about 10 square yards). Each pedon approximates a polygonal configuration in the horizontal, has a depth equivalent to the soil profile (Figure 4.6), and is the smallest volume that is representative of the total soil.

A *polypedon* is made up of two or more contiguous pedons significantly equivalent in characteristics to be identified as being the same

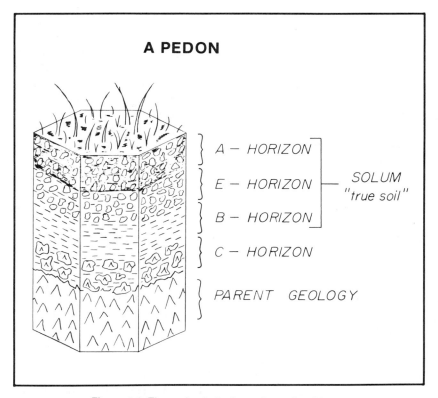

Figure 4.6 The pedon in its three-dimensional form.

soil. The polypedon concept is consistent with the term soil *individual,* and for classification purposes, polypedons of like traits are called a *soil* series. The soil series definition serves as the primary basis for determining units of classification in the U. S. Comprehensive Soil Classification System, frequently referred to as the "Soil Taxonomy."

Figure 4.7 identifies the soil series as related to landscape position in the North Carolina Piedmont. The reader may find similarity in the format of this diagram and Figure 4.3, which illustrates the catena concept. Indeed, a catena is represented in Figure 4.7 following the sequence Cecil-Vance-Wilkes-Monacan. Each of the foregoing is also a soil series, having unique attributes that are repeated within landforms subject to parallel sets of pedogenic processes.

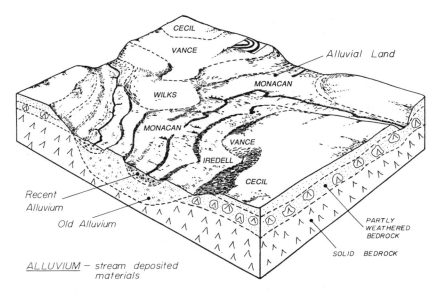

Figure 4.7 Soil Series as they relate to landscape position.

Soil Classification

5

Soil characteristics vary through time and space. A complete analysis of all the constituents of several soil profiles from a farmer's single field, in fact, would reveal that each pedon is unique—no two soils are exactly alike. Just as there are some (however small) differences in even the most seemingly perfectly matched identical twins, pedons vary over the earth's surface. On the other hand, soils that are affected by similar environmental controls do tend to exhibit common properties related to their relatively uniform genetic background, at the same time showing differences derived from occupying a particular niche in a particular region.

Classification of soils is designed to satisfy practical needs. In a classical treatise, Baldwin, Kellogg, and Thorp (1938, 979) state, "Man has a passion for classifying everything. There is reason for this; the world is so complex that we could not understand it at all unless we classified like things together. Just as plants, insects, birds, minerals, and thousands of other things are classified, so are soils."

In relation to soils, the "passion" to classify is not a recent phenomenon associated with modern technology and mechanized cultivation, but has probably been a conscious factor since the beginning of agriculture. There is evidence that the Chinese recognized different kinds of soils, and named them, over 4,000 years ago. This early classification was based mainly on soil color and structure. Within the last 150 years or so, data from soil analyses have become more abundant and have resulted in more sophisticated soil classification methods. Seeking the causes behind the recurrence of regionally distributed soil features, investigators in the first half of the nineteenth century attempted to categorize soils on the traits of parent material—they assumed that geology was the primary factor in determining soil composition. This approach applied well to young soils, but provided unsatisfactory results for many mature sites where climate and vegetation had substantially altered the parent material.

In the latter half of the nineteenth century, a fresh approach to pedology was initiated by the Russian scientist V. V. Dokuchaiev. Soil classification now focused on the concept of soil as an independent natural body exhibiting distinct characteristics that resulted from the interactions of climate, parent material, flora and fauna, geomorphic factors, and time.

The Dokuchaiev school, founded about 1870, has had a profound impact on more recent pedologic research. The publications of his first students mark the inception of modern soil science. No longer was soil viewed simply as being produced by the physical and chemical alteration of its parent material. Rather, it was considered to be the evolutionary product of a dynamic system (1) whose current status was the result of several energy inputs; (2) with attributes that vary through time; (3) whose developmental factors could be identified; and (4) that, due to the varying degree of dominance of specific pedogenic agents, had recognizable variations that could be displayed on maps.

Research now focused on determining general soil characteristics and their areal distribution. Several genetic soil classification plans that came from those early investigations stressed the interrelated role of climate and vegetation as the primary pedogenetic agent.

Other countries had very little knowledge of Russian advances in soil investigation, however, until the 1914 publication in Berlin of K. D. Glinka's *Die Typen der Bodenbildung, Ihre Klassification und Geographische Verbreitung.* (Glinka was a student of Dokuchaiev's.) C. F. Marbut became familiar with Glinka's writings, and through his influence as chief of the U. S. Soil Survey, soil identification and mapping in the United States turned from a geologic orientation to one "based primarily on soil profile studies and their genetic implications" (Kellogg, 1963, 3). Aligning with the Dokuchaiev philosophy, Marbut developed a genetic soil classification system that utilized considerable Russian terminology (Table 5.1).[1]

Marbut stressed the distinction between dynamic (climate and biologic) and passive (parent material, topographic position, and time) soil development factors. He also distinguished soil proper (the *solum*) from its underlying geologic material (C horizon) and recognized the geographic expression of soils in the emphasis he placed on mature (zonal) soils—

---

[1] Marbut proposed an initial system of soil classification in 1927 that gave strong emphasis to mature soils. He continued to refine the system until it progressed to his 1936 version, which is provided in Table 5.1.

Table 5.1 Soil Categories

Category VI	Pedalfers (VI-1)	Pedocals (VI-2)
Category V	Soils from mechanically comminuted materials Soils from siallitic decomposition products Soils from allitic decomposition products	Soils from mechanically comminuted materials
Category IV	Tundra Podzols Gray-brown podzolic soils Red soils Yellow soils Prairie soils Lateritic soils Laterite soils	Chernozems Dark-brown soils Brown soils Gray soils Pedocalcic soils of Arctic and tropical regions
Category III	Groups of mature but related soil seris Swamp soils Glei soils Rendzinas Alluvial soils Immature soils on slopes Salty soils Alkali soils Peat soils	Groups of mature but related soil series Swamp soils Glei soils Rendzinas Alluvial soils Immature soils on slopes Salty soils Alkali soils Peat soils
Category II	Soil series	Soil series
Category I	Soil units, or types	Soil units, or types

category IV in his classification system. These were considered to be "great soil groups" (Figure 5.1) and consisted of hundreds of *soil series* that differed from one another due to variation in parent material, relief, and age, yet all exhibited the same general sort of profile.[2] The definition of soil series places considerable emphasis on properties of importance to soil behavior. Ranges in properties of series are narrow in highly

---

[2] *Soil series,* the basic unit of soil classification, being the subdivision of a family and consisting of soils that are essentially alike in all major profile characteristics, except the texture of the A horizon.

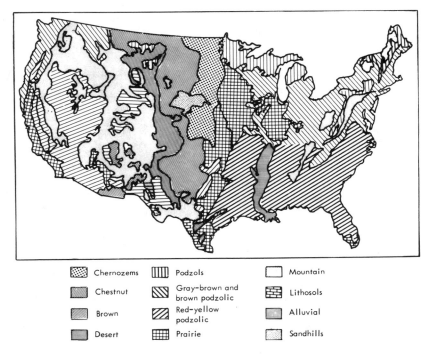

Chernozems   Podzols   Mountain

Chestnut   Gray-brown and brown podzolic   Lithosols

Brown   Red-yellow podzolic   Alluvial

Desert   Prairie   Sandhills

Figure 5.1 Great Soil Groups of the United States according to the Marbut classification scheme.

productive soils, and broad in soils of low productivity or unsuited to farming.

A demand for major change in classification criteria came about during World War II when an increased need for food and fiber motivated worldwide government-sponsored soil research and resulted in the identification of numerous soils previously unclassified. Immediately after the war, serious attempts were made to incorporate known soil series, including those newly discovered, into the Marbut classification scheme. These efforts failed even after attempts to revise definitions of the great soil groups (Table 5.2 and Figure 5.1). The well-known soil scientist C. E. Kellogg (1963, 3) remarked that most systems of soil classification in general use as late as 1950 had serious faults. Several emphasized pedogenesis and required assumptions regarding the soil's character under virgin conditions. Soils that had been severely eroded, drastically reworked, or made from transported materials were difficult to place within such genetic frameworks. A new classification method seemed necessary,

Table 5.2   Soil Classification in the Higher Categories (Revised, 1949)

Order	Suborder	Great Soil Groups
Zonal soils	1. Soils of the cold zone	Tundra soils
	2. Light-colored soils of arid regions	Desert soils
		Red Desert soils
		Sierozem
		Brown soils
		Reddish-Brown soils
	3. Dark-colored soils of semi-arid, subhumid, and humid grasslands	Chestnut soils
		Reddish Chestnut soils
		Chernozem soils
		Prairie soils
		Reddish Prairie soils
	4. Soils of the forest-grassland transition	Degraded Cherozem
		Noncalcic Brown or Shantung Brown soils
	5. Light-colored podzolized soils of the timbered regions	Podzol soils
		Gray wooded, or Gray Podzolic soils
		Brown Podzolic soils
		Gray-Brown Podzolic soils
		Red-Yellow Podzolic soils
	6. Lateritic soils of forested-warm-temperate and tropical regions	Reddish-Brown Lateritic soils
		Yellowish-Brown Lateritic soils
		Laterite soils
Intrazonal soils	1. Halomorphic (saline and alkali) soils of imperfectly drained arid regions and littoral deposits	Solonchak, or saline soils
		Solonetz soils
		Soloth soils
	2. Hydromorphic soils of marshes, swamp, seep areas, and flats	Humic Gley soils (includes Wiesenboden)
		Alpine Meadow soils
		Bog soils
		Half-Bog soils
		Low-Humic Gley soils
		Planosols
		Ground-Water Podzol soils
		Ground-Water Laterite soils
	3. Calcimorphic soils	Brown-Forest soils (Braunerde). Rendzina soils
Azonal soils		Lithosols
		Regosols (includes Dry Sands)
		Alluvial soils

one that would avoid past weaknesses and in which similar soils, whether cultivated or virgin, would be placed within the same taxon (group or entity).

U. S. COMPREHENSIVE SOIL CLASSIFICATION SYSTEM

During the 1950s the Soil Survey Staff of the U. S. Department of Agriculture, under the chairmanship of Guy D. Smith, assumed responsibility for developing the U. S. Comprehensive Soil Classification System as a direct response to recognized weaknesses of methods currently in use. The comprehensive methodology was completely new above the level of soil series, and was developed by a successive set of *approximations,* each of which was circulated to professional soil scientists for test application and criticism. These approximations, with successive supplements and revisions, became the body of data from which the current Comprehensive Classification System was derived and subsequently published in the USDA's *Soil Taxonomy* (1975).

The *taxa* above the soil series level were given completely new names,[3] because much of the terminology previously used to describe soil characteristics had been redefined many times and had held different meanings at different periods and in different parts of the world. Although the focus of the present system is based on properties of soils as they exist today (eliminating shortcomings of previous systems that identified soil on the basis of the properties it would possess under virgin conditions), it does not completely ignore genesis. Since soil properties are directly related to soil development, genesis is implicitly considered in the current methodology.

Six levels of generalization have been established, and soils are placed in a common taxon only if there is evidence of one or more dominant processes sufficiently affecting the soil to produce common diagnostic horizons or features.[4] At the *order level,* only a few similarities are present. In the lowest category (soil series), there is relatively complete homogeneity in both soil features and their genesis. The criteria for separation of the various taxa is briefly summarized below:

---

[3] Individuals or items in any type of population may be grouped according to common characteristics or properties. A group so formed is known as a class or *taxon* (plural, *taxa*).

[4] Each soil exhibits distinct features that are characteristic of the environment in which it has formed. These features are said to be "diagnostic" in that they distinguish the soils in one taxon from those in another.

I. Order level: A given soil must possess common properties indicating similarity in kind and strength of pedogenic processes, such as the presence or absence of major diagnostic horizons.

II. Suborder level: This class has genetic homogeneity. Subdivision of orders is according to the presence or absence of properties associated with wetness, soil moisture regimes, major parent material, and vegetational effects, as indicated by key properties.

III. Great Group level: Differentiation is based upon similar kind, arrangement, and degree of expression of horizons, with emphasis on the upper sequum; base status; soil temperature and moisture regimes; and the presence or absence of diagnostic layers, such as plinthite, fragipan, and others.

IV. Subgroup level: Provides for the central concept of the great group and for properties indicating intergradations to other great groups, suborders, and orders, and to extragradation to "not soil."

V. Family level: This class identifies features of importance to plant root growth, such as broad textural characteristics, mineralogical composition, and soil temperature and reaction.

VI. Series level: Soils are separated on the kind and arrangement of horizons; color, texture, structure, consistency, and reaction of horizons; and the chemical and mineralogical properties of the horizons.

The order level represents the greatest degree of soil generalization and consists of ten categories, each possessing an identfying name ending in *sol* (Latin *solum,* soil). They are listed in Table 5.3 along with their root word derivations, percentage of areal dominance, and approximate equivalents in the latest revised Marbut System. (Steila, 1976, 70)

Each of the mineral soils at the order level occupies a position within a hierarchy that is largely based upon *degree of weathering,* that is, degree of mineral alteration and profile development (Figure 5.2). Their global distribution is shown in Figure 5.3. The *Entisols* are youthful soils characterized as lacking natural genetic horizons or at best having only the beginnings of such. *Vertisols* are soils containing large amounts of montmorillinite (shrinking and swelling) clays. Because of alternate shrinking and swelling this soil will become "inverted" with time. *Inceptisols* are soils that are beginning to show development of genetic horizons. They do not have evidence of extreme weathering and are not sufficently developed to be classified in one of the seven remaining orders.

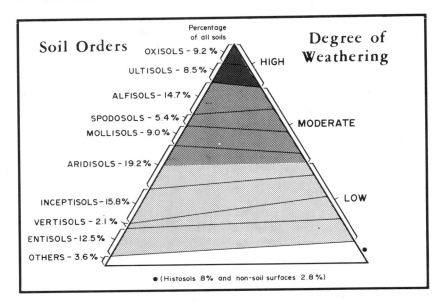

Figure 5.2 Relationship between soil orders and intensity of weathering. (The percentages provide the approximate areal extent of each Soil Order.)

Aridisols are light-colored soils of desert climatic regimes. *Mollisols* are typically found in grassland environments. They have a soft, thick, dark-colored surface horizon. *Spodosols* are primarily located in cool and humid forested regions. They contain a subsurface horizon wherein amorphous organic matter and aluminum with or without iron have accumulated. *Alfisols* are strongly weathered soils in which translocation of aluminum and iron has taken place; yet they retain a relatively high base saturation. *Ultisols* are extremely weathered soils with low retention of bases. *Oxisols* are found within the tropics. They contain subsurface horizons consisting of a mixture of hydrated oxides of iron and/or aluminum and 1:1 lattice clays. *Histosols* represent the organic soils and are composed primarily of vegetative debris in various stages of decomposition.

As a result of macroscale pedogenesis, there is regional homogeneity in diagnostic soil characteristics. A close examination of Figure 5.3 reveals broad patterns in the distribution of soil orders that can be readily associated with relatively extensive climatic and vegetative realms. Yet each order comprises hundreds of subtypes, each of which has attributes associated with its unique intraregional environment. The taxa of the lower categories provide an orderly identification of such differences.

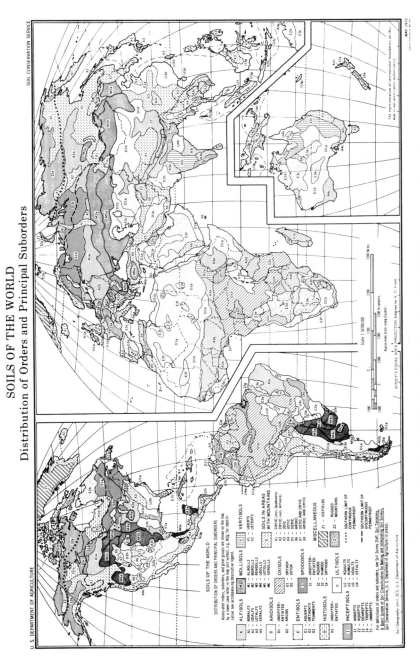

Figure 5.3 World distribution of soils (courtesy of the Soil Conservation Service.)

75

Table 5.3 *Soil Orders, Name Derivation, Areal Significance, and Marbut Equivalents*

Soil Order	Derivation of root word	% of Total world soils [a]	Rank (total area)	Marbut equivalents
Entisols	Recent	12.5	4	Azonal soils, some Low Humic Glay
Vertisols	L. *verto*, to turn	2.1	9	Grumusols
Inceptisols	L. *inceptum*, inception, beginning	15.8	2	Ando, Sol Brun Acide, some Brown Forest, Low Humic Gley, Humic Gley
Aridisols	L. *aridus*, dry	19.2	1	Desert, Reddish Desert, Sierozem, Solonchak, some Brown and Reddish Brown soils, associated Solonetz
Mollisols	L. *mollis*, soft	9.0	6	Chestnut, Chernozem, Brunizem, Rendzina, some Brown, Brown Forest, associated Humic Gley, and Solonetz
Spodosols	Gr. *spodos*, wood ash	5.4	8	Podzols, Brown Podzolic, Ground-water Podzols
Alfisols	*Alf* combined from aluminum and iron	14.7	3	Gray-Brown Podzolic, Gray Wooded, Noncalcic Brown, Degraded Chernozem, associated Planosols and Half-Bog
Ultisols	L. *ultimos*, ultimate	8.5	7	Red-Yellow Podzolic, Reddish-Brown Lateritic, associated Planosols, and some Half-Bogs
Oxisols	*Oxi* from oxide	9.2	5	Laterite soils, Latosols
Histisols	Gr. *histos*, tissue	0.8	10	Bog soils

[a] An additional 2.8 percent of the world total includes ice fields, unclassified lands, and others.

Table 5.4 Formative Elements in Names of Suborders

Formative element	Derivation of formative element	Mnemonicon	Connotation of formative element
alb	L. *albus*, white	albino	Presence of albic horizon (a bleached eluvial horizon)
and	Modified from *Ando*	Ando	Andolike
aqu	L. *aqua*, water	aquarium	Characteristics associated with wetness
ar	L. *arare*, to plow	arable	Mixed horizons
arg	Modified from argillic horizon; L. *argilla*, white clay	argillite	Presence of argillic horizon (a horizon with illuvial clay)
bor	Gr. *boreas*, northern	boreal	Cool
ferr	L. *ferrum*, iron	ferruginous	Presence of iron
fibr	L. *fibra*, fiber	fibrous	Least decomposed stage
fluv	L. *fluvius*, river	fluvial	Flood plains
hem	Gr. *hemi*, half	hemisphere	Intermediate stage of decomposition
hum	L. *humus*, earth	humus	Presence of organic matter
lept	Gr. *leptos*, thin	leptometer	Thin horizon
ochr	Gr. base of *ochros*, pale	ocher	Presence of ochric epipedon (a light-colored surface)
orth	Gr. *orthos*, true	orthophonic	The common ones
plag	Modified from Ger. *plaggen*, sod		Presence of plaggen epipedon
psamm	Gr. *psammos*, sand	psammite	Sand textures
rend	Modified from *Rendzina*	Rendzina	Rendzinalike
sapr	Gr. *sapros*, rotten	saprophyte	Most decomposed stage
torr	L. *torridus*, hot and dry	torrid	Usually dry
trop	Modified from Gr. *tropikos*, of the solstice	tropical	Continually warm
ud	L. *udus*, humid	udometer	Of humid climates
umbr	L. *umbra*, shade	umbrella	Presence of umbric epipedon (a dark colored surface)
ust	L. *ustus*, burnt	combustion	Of dry climates, usually hot in summer
xer	Gr. *xeros*, dry	xerophyte	Annual dry season

Table 5.5   Formative Elements in Names of Great Groups

Formative element	Derivation of formative element	Mnemonicon	Connotation of formative element
acr	Modified from Gr. *akros*, at the end	acolith	Extreme weathering
agr	L. *ager*, field	agriculture	An agric horizon
alb	L. *albus*, white	albino	An albic horizon
and	Modified from *Ando*	Ando	Andolike
anthr	Gr. *anthropos*, man	anthropology	An anthropic epipedon
aqu	L. *aqua*, water	aquarium	Characteristic associated with wetness
arg	Modified from argillic horizon: L. *argilla*, white clay	argillite	An argillic horizon
calc	L. *calcis*, lime	calcium	A calcic horizon
camb	L. *cambiare*, to exchange	change	A cambic horizon
chrom	Gr. *chroma*, color	chroma	High chroma
cry	Gr. *kryos*, coldness	crystal	Cold
dur	L. *durus*, hard	durable	A duripan
dystr	Modified from Gr. *dys*, ill, *dystrophic*, infertile	dystrophic	Low base saturation
eutr	Modified from Gr. *eu*, good; *eutrophic*, fertile	eutrophic	High base saturation
ferr	L. *ferrum*, iron	ferric	Presence of iron
frag	Modified from L. *fragilis*, brittle	fragile	Presence of fragipan
fragloss	Compound of *fra(g)* and *gloss*		See the formative elements *frag* and *gloss*
gibbs	Modified from *gibbsite*	gibbsite	Presence of gibbsite
gloss	Gr. *glossa*, tongue	glossary	Tongued
hal	Gr. *hals*, salt	halophyte	Salty
hapl	Gr. *haplous*, simple	haploid	Minimum horizon
hum	L. *humus*, earth	humus	Presence of humus
hydr	Gr. *hydr*, water	hydrophobia	Presence of water
hyp	Gr. *hypnon*, moss	hypnum	Presence of hypnum moss
luo, lu	Gr. *louo*, to wash	ablution	Illuvial
moll	L. *mollis*, soft	mollify	Presence of mollic epipedon
nadur	Compound of *na(tr)*, and *dur*		
natr	Modified from *natrium*, sodium		Presence of natric horizon
ochr	Gr. base of *ochros*, pale	ocher	Presenence of ochric epipedon

Table 5.5   (Continued)

Formative element	Derivation of formative element	Mnemonicon	Connotation of formative element
pale	Gr. *paleos*, old	paleosol	Old development
pell	Gr. *pellos*, dusky		Low chroma
plac	Gr. base of *plax*, flat stone		Presence of a thin pan
plag	Modified from Ger. *plaggen*, sod		Presence of plaggen horizon
plinth	Gr. *plinthos*, brick		Presence of plinthite
quartz	Ger. *quartz*, quartz	quartz	High quartz content
rend	Modified from *Rendzina*	Rendzina	Rendzinalike
rhod	Gr. base of *rhodon*, rose	rhododendron	Dark-red colors
sal	L. base of *sal*, salt	saline	Presence of salic horizon
sider	Gr. *sideros*, iron	siderite	Presence of free iron oxides
sombr	Fr. *sombre*, dark	somber	A dark horizon
sphagno	Gr. *sphagnos*, bog	sphagnum moss	Presence of sphagnum moss
torr	L. *torridus*, hot and dry	torrid	Usually dry
trop	Modified from Gr. *tropikos*, of the solstice	tropical	Continually warm
ud	L. *udus*, humid	udometer	Of humid climates
umbr	L. base of *umbra*, shade	umbrella	Presence of umbric epipedon
ust	L. base of *ustus*, burnt	combustion	Dry climate, usually hot in summer
verm	L. base of *vermes*, worm	vermiform	Wormy, or mixed by animals
vitr	L. *vitrum*, glass	vitreous	Presence of glass
xer	Gr. *xeros*, dry	xerophyte	Annual dry season

The division of soil orders into suborders is based upon chemical and/or physical properties that indicate drainage conditions or upon genetic differences due to climate and vegetation. The name of the suborder is made up of two syllables. The last syllable always indicates the order, for example, *Ult* for Ultisol. The prefix identifies a characteristic unique to a particular suborder, such as *aqu* for aqua (Latin, *aqua,* water), meaning wet. Hence, an Ultisol exhibiting signs of wetness would be classified in the suborder *Aquult.* Likewise, an Entisol with a sandy surface horizon would be considered a Psamment (Greek *psammos,* sand). The formative elements identifying factors of importance at the suborder level are listed in Table 5.4.

Suborders are separated into *great groups* on the basis of kind and array of diagnostic horizons. The name of the great group indicates the type of features present by prefixing one or more formative elements onto the correct suborder name. For example, a *Plinthaquult* is a wet Ultisol that has plinthite within the profile, and a *Quartzipsamment* is a sandy Entisol comprised primarily of quartz crystals. (See Table 5.5 for formative elements used to identify great groups.)

Great groups are subdivided into *subgroup* categories that indicate to what extent the central concept of the great group is expressed. A typic *Plinthaquult* specifies a soil typical of the great group Plinthaquult.

The lowest taxonomic categories are *family* and series. The families stress features of importance to plant growth, such as texture, mineralogy, reaction class, and temperature. The series identify the individual soil, named after a natural feature or place where the soil was first discovered. A complete soil classification for the Pantego series, along with its profile description, follows.

PANTEGO SERIES

The Pantego series is a member of the fine-loamy, siliceous, thermic family of Umbric Paleaquults (Table 5.6). These soils have black or very dark gray, fine, sandy, loam A horizons and grayish, sandy, clay loam Bt horizons.

Table 5.6   *Profile Description of an Umbric Paleaquult*

Horizon [a]	Depth (Cm)	Typifying pedon: Pantego fine, sandy loam—Cultivated field (Colors are for moist soil)
Ap	0-25	Black (10YR 2/1) [b] fine sandy loam; weak, fine-granular structure; very friable; many fine roots; very strongly acid; gradual wavy boundary (15 cm to 30 cm thick).
A12	25-45	Very dark gray (10YR 3/1) fine sandy loam; weak, fine-granular structure; friable; very strongly acid; clear smooth boundary (10 cm to 20 cm thick).
B21tg	45-70	Very dark gray (10YR 3/1) sandy clay loam; weak, fine, subangular blocky structure; friable; patchy clay films on ped faces and in pores; very strongly acid; gradual wavy boundary (22 cm to 44 cm thick).
B22tg	70-106	Gray (10YR 5/1) sandy clay loam; few fine and medium distinct mottles of brownish yellow (10YR 6/6); weak, fine and medium subangular blocky structure; friable; slightly sticky; patchy clay films on ped faces; very strongly acid; gradual smooth boundary (22 cm to 50 cm thick).
B23tg	106-140	Gray (10YR 6/1) sandy clay loam; few medium and coarse distinct mottles of yellowish brown (10YR 5/6); weak, fine subangular blocky structure; friable; slightly sticky; patchy clay films on ped faces; very strongly acid; gradual wavy boundary (30 cm to 45 cm thick).
B3g	140-165	Gray (10YR 6/1) sandy clay loam; weak coarse subangular blocky structure; friable; patchy clay films on ped faces; very strongly acid.

[a] For a description of symbols used to describe the characteristics of a soil profile, refer to Appendix II. The reader should make such reference each time a soil profile description occurs in the remainder of the text.

[b] For a description of symbols used to describe soil color, refer to Appendix III.

Entisols, Vertisols, and Inceptisols

6

The Entisols, Vertisols, and Inceptisols are mineral soils that lack sufficient horizon development and/or chemical alteration to produce the mature soil characteristics associated with their regional bioclimatic environment. In most instances, these soils extend beyond the limits of any particular climatic region or ecosystem. An example is soil formed in *alluvium* (stream-deposited sediments), like those in valleys of the Mississippi River and its tributaries. Such alluvial soils can occur under practically any thermal or moisture regime and under widely varying vegetative covers.

The primary difference among these three soil orders is the degree to which they may approach the status of a mature profile. Entisols exhibit little or no evidence of pedogenic horizons. The Vertisols are dominated by clays with a high shrink-swell capacity and exhibit some mineral alteration. Inceptisols have weakly developed pedogenic features attributable to their soil-forming environment.

ENTISOLS

Soil-forming processes in the Entisols either have not been operative for a sufficient period of time or some aspect of the physical environment has prevented complete development of pedogenic horizons. These are true soils and should not be confused with recently weathered parent material incapable of supporting plant life. Various factors are responsible for the lack of horizon development or mineral alteration in Entisols. In some, time has been a dominating element. Such Entisols are found on newly exposed surface deposits that have not been in place long enough for pedogenesis to operate to its fullest. Some may be found on steep, actively eroding slopes, others on glacial outwash-plains, and still others on floodplains that receive new deposits of alluvium at frequent intervals. Some Entisols, however, are very old. Consisting primarily of quartz or other minerals that do not alter readily, they simply do not form horizons. Entisols may exist in almost any moisture or temperature region, on any type of parent material, and under any form of vegetation (Figure 6.1).

Most maps illustrating the areal distribution of soils (such as Figures 5.1, 5.3, and 6.1) are based upon broad generalizations that account solely for the most dominant soil in any particular area. For example, suppose a given county has the following percentage of soils, by land area: 35 percent Ultisols, 30 percent Entisols, 25 percent Inceptisols, and 10 percent Alfisols. On maps where soil is represented by a number of discrete classes, the cartographer is limited to one choice. Thus, our hypothetical county would be classified and mapped as having Ultisols. Note that in this process much valuable information is lost, and the procedure can mislead the map user by not providing insight for 65 percent of the county's soils.

To compensate for the "mapping problem," Gersmehl (1977) constructed dot distribution maps for each soil order within the contiguous

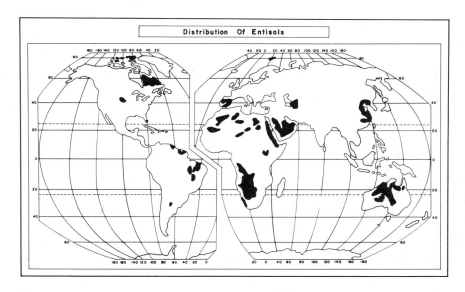

Figure 6.1 World distribution of Entisols.

United States. Each dot represents 1,000 square kilometers of a particular soil's occurrence as recorded in sampled county soil surveys, but does not represent a continuous, uniform expanse of a given soil. Figure 6.2 is a reproduction of his Entisol map. When used in conjunction with maps of areal dominance, such as Figures 5.3 and 6.1, such a map will greatly add to one's knowledge of a soil's distribution. (Unfortunately such maps are not available at a global scale.) In the remaining chapters, both map forms are incorporated into the text.

There are five suborders of Entisols included in Figure 6.3, as follows:

1. *Aquents* (Latin *aqua,* water) are the wet Entisols. They are commonly found in tidal marshes, in deltas on margins of lakes where the soil is continuously saturated with water, in floodplains of streams where soil is saturated at some time of the year, or in very wet sandy deposits. These soils are bluish or gray and mottled. Temperature does not restrict the occurrence of Aquents, except in areas constantly below freezing. The soil moisture regime is primarily a reducing one, virtually free of dissolved oxygen due to lengthy saturation by ground water or its capillary fringe. Thus, Aquents are generally found in recent, usually water-deposited sediments, but may support any form of vegetation that will tolerate prolonged periods of wetness.

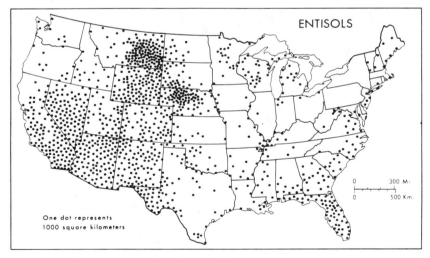

Figure 6.2 Entisols of the United States. (From Philip J. Gersmehl, "Soil Taxonomy and Mapping," *Annals of the Association of American Geographers,* 67, September 1977, p. 423. By permission of the Association of American Geographers.)

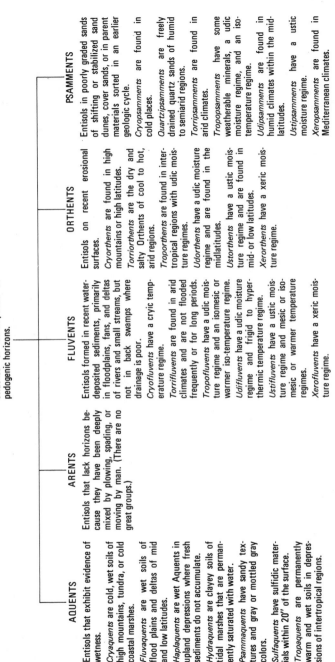

ENTISOLS

Soils that have little or no evidence of development of pedogenic horizons.

AQUENTS

Entisols that exhibit evidence of wetness.

Cryaquents are cold, wet soils of high mountains, tundra, or cold coastal marshes.

Fluvaquents are wet soils of flood plains and deltas of mid and low latitudes.

Haplaquents are wet Aquents in upland depressions where fresh sediments do not accumulate.

Hydraquents are clayey soils of tidal marshes that are permanently saturated with water.

Psammaquents have sandy textures and gray or mottled gray colors.

Sulfaquents have sulfidic materials within 20" of the surface.

Tropaquents are permanently warm and wet soils in depressions of intertropical regions.

ARENTS

Entisols that lack horizons because they have been deeply mixed by plowing, spading, or moving by man. (There are no great groups.)

FLUVENTS

Entisols formed in recent water-deposited sediments, primarily in floodplains, fans, and deltas of rivers and small streams, but not in back swamps where drainage is poor.

Cryofluvents have a cryic temperature regime.

Torrifluvents are found in arid climates and are not flooded frequently or for long periods.

Tropofluvents have a udic moisture regime and an isomesic or warmer iso-temperature regime.

Udifluvents have a udic moisture regime and frigid to hyperthermic temperature regime.

Ustifluvents have a ustic moisture regime and mesic or iso-mesic or warmer temperature regimes.

Xerofluvents have a xeric moisture regime.

ORTHENTS

Entisols on recent erosional surfaces.

Cryorthents are found in high mountains or high latitudes.

Torriorthents are the dry and salty Orthents of cool to hot, arid regions.

Troporthents are found in intertropical regions with udic moisture regimes.

Udorthents have a udic moisture regime and are found in the midlatitudes.

Ustorthents have a ustic moisture regime and are found in mid- or low latitudes.

Xerorthents have a xeric moisture regime.

PSAMMENTS

Entisols in poorly graded sands of shifting or stabilized sand dunes, cover sands, or in parent materials sorted in an earlier geologic cycle.

Cryopsamments are found in cold places.

Quartzipsamments are freely drained quartz sands of humid to semiarid regions.

Torripsamments are found in arid climates.

Tropopsamments have some weatherable minerals, a udic moisture regime, and an iso-temperature regime.

Udipsamments are found in humid climates within the mid-latitudes.

Ustipsamments have a ustic moisture regime.

Xeropsamments are found in Mediterranean climates.

Figure 6.3 Suborders and Great Groups of the Soil Order Entisol.

2. *Arents* (Latin *arare*, to plow) are Entisols that lack horizons, normally because of human interference. They have been deeply mixed by plowing, spading, or moving. Arents may contain fragments that can be identified as parts of former diagnostic horizons, yet the fragments are no longer continuous. Instead, remnants are scattered throughout the solum and mixed with other horizons. Some of these soils are the result of deliberate human attempts to either modify soil or to break up and remove restrictive pans;[1] in other instances they result from cut-and-fill operations intended to reshape the surface. An uncommon few are formed from the natural effect of mass movement in earth slides. Arents are not extensively developed, and their characteristics vary from place to place. Powerful modern machinery seems to be increasing their acreage, and they are likely to become more significant in the future.

3. *Fluvents* (Latin *fluvius,* river) are brownish to reddish Entisols. They have formed in recent water-deposited sediments—primarily floodplains, fans, and deltas of rivers and small streams, but not in back swamps where drainage is poor. Although the time span could, in specific instances, be as long as a few hundred years in humid regions, and even longer in arid areas, the age of sediments in which these soils form is usually very young, a few years or decades. Under normal conditions Fluvents are flooded frequently, and deposited materials show signs of stratification, that is, layers of a given texture alternate with layers of other textures. Most alluvial sediments, coming from eroding surfaces or stream banks, include appreciable amounts of organic carbon that are dominantly associated with the clay fraction. Hence, clayey or loamy strata usually have more organic carbon than strata that either overlie or underlie them and that are more sandy. The percentages of organic carbon, therefore, decrease irregularly with depth if the materials are stratified. If textures are homogeneous, on the other hand, the organic carbon content will decrease regularly with depth. These soils do not occur under any specific vegetation and may be found in any moisture or thermal regime, except those that are subject to temperatures constantly below freezing.

4. *Orthents* (Greek *orthos*, true) are Entisols occurring primarily on recently eroded surfaces that have been created by geologic factors or produced by cultivation. The basic requirement is that any former existing soil has been either completely removed or so truncated that the diagnostic horizons typical of all orders other than Entisols are absent. In some cases remnants of indurated diagnostic horizons, such as ironstone that once may have been plinthite, may be present if exposed at the surface.[2] Such

---

[1] A *pan* is a layer or horizon within a soil that is firmly compacted and/or is very rich in clay.

[2] Plinthite is discussed in detail in Chapter 11.

formations normally support only scattered plants. If they do not support plants, they are considered to be rock rather than soil. A few Orthents occur in recent loamy or fine wind-deposited sediments, glacial deposits, debris from recent landslides and mudflows, and in recent sandy alluvium.

5. *Psamments* (Greek *psammos,* sand) are sandy Entisols that lack pedogenic horizons. They include sandy dunes, cover sands, and sandy parent material produced in an earlier geologic cycle. Others are found in sands that have been sorted by water, natural levees or beaches. Occurring under any climate or vegetation, they may be located on surfaces of virtually any age. The older Psamments are usually dominated by quartz sand and cannot form subsurface diagnostic horizons that involve the accumulation of clays and sesquioxides.

A profile description of a Psamment is shown in Table 6.1. This soil is found in Calhoun County, Florida, and is a member of the thermic, coated family of Typic Quartzipsamments. These soils have thin, very dark grayish brown, sandy A horizons and yellowish brown, sandy C horizons. Note that no E or B horizon, typical of mature soils, is present.

Table 6.1   Profile Description of a Psamment

Horizon	Depth (Cm)	Typifying pedon: Lakeland Sand
A	0-8	Very dark grayish brown (10YR 3/2 crushed) sand; single-grained; loose; moisture of very dark grayish brown (10YR 3/2) organic matter and clean, uncoated, white (10YR 8/1) sand grains; common fine and medium roots; strongly acid; clear wavy boundary (5 cm to 20 cm thick).
C1	8-25	Yellowish brown (10YR 5/4) sand; common splotches of yellowish brown (10YR 5/6); single-grained; loose; common and fine medium roots; few uncoated sand grains; strongly acid; gradual wavy boundary (10 cm to 25 cm thick).
C2	25-108	Yellowish brown (10YR 5/8) sand; single grained; loose; few fine roots; few uncoated sand grains; strongly acid; gradual wavy boundary (75 cm to 125 cm thick).
C3	108-160	Yellowish brown (10YR 5/8) sand; few medium faint splotches of very pale brown (10YR 7/3, 7/4); single-grained; loose; many uncoated sand grains; strongly acid; gradual wavy boundary (45 cm to 75 cm thick).
C4	160-225	Very pale brown (10YR 7/4) sand; few medium distinct yellowish red (5YR 4/8) mottles; single-grained; loose; many uncoated sand grains; strongly acid.

VERTISOLS

Vertisols are clayey soils that have deep, wide cracks during periods of moisture deficiency. The clays making up these soils swell upon wetting and shrink when dried. In the United States, the dominant clay mineralogy is montmorillinite, although in other parts of the world illite or a mixed mineralogy may produce the same characteristics.

Formation of Vertisols involves the production of 2:1 layer-lattice, expanding clays by weathering activity within a climatic realm that experiences strong periodic contrasts in its moisture supply (Figures 6.4 and 6.5).[3] The source of the clay may be argillaceous sedimentary materials or the result of the chemical alteration of basic igneous products. In either case, maintaining the soil's shrink-swell properties depends upon whether further weathering destroys the clay minerals' expanding character.

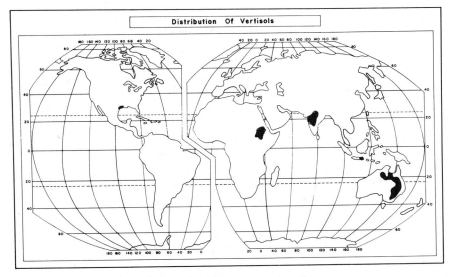

Distribution Of Vertisols

Figure 6.4 World distribution of Vertisols.

---

[3] In arid regions Vertisols commonly develop in closed depressions or playas that are occasionally flooded or in fine-textured materials in areas that have infrequent heavy showers. In other regions seasonal moisture changes are chiefly the result of excesses of evapotranspiration over precipitation in one season, followed by an excess of precipitation in the next.

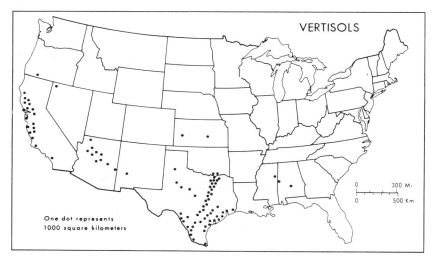

Figure 6.5 Vertisols of the United States. (From Philip J. Gersmehl, "Soil Taxonomy and Mapping," *Annals of the Association of American Geographers,* 67, September 1977, p. 424. By permission of the Association of American Geographers.)

With partial drying, the clays shrink and produce wide, deep cracks. These open cracks may be more than 2.5 centimeters wide to a depth of 50 centimeters. In certain areas, cracks have been known to extend to depths in excess of 1 meter. By definition, an open crack is considered a separation between very coarse prisms or polyhedrons. The open cracks permit surface materials that are displaced by precipitation, animals, or the like to fall to lower depths of the solum within the confines of the separations. When rain occurs, water runs into the cracks readily; hence the soil is remoistened both from above and below. With moistening, the clays expand and the cracks close, trapping the displaced particles at lower levels (Figure 6.6). As the profile moistens from below, the clays in lower horizons tend to swell before those above them. This results in one part of the soil moving against another as pressure increases due to an increased volume of material in the lower portion of the solum. Pressure is exerted in all directions, but the soil is capable of moving only horizontally or upward. It is thought that this process is the probable means of producing *slickensides* and *gilgai.* Slickensides are polished and grooved surfaces in the soil made by one mass of material sliding

The Soil Survey staff recognizes four suborders of Vertisols (Figure 6.7), as follows:

1. *Torrerts* (Latin *torridus,* hot, dry) are the Vertisols of arid climatic regions. They have cracks that can remain open the entire year, although

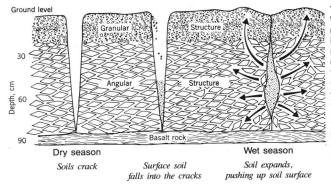

The formation of Vertisols. From left to right: (1) Cracks develop in dry season. (2) Loose material falls into cracks. (3) Wetting of the soil in the wet season causes expansion and movement of soil in the lower part of the soil to produce angular or wedge-shaped peds with shiny surfaces (slickensides) and a microrelief called gilgai. (Adapted from Boul, 1966.)

Figure 6.6 Vertisol Development. (From H.D. Foth, *Fundamentals of Soil Science,* 7th ed., New York: John Wiley and Sons, Publishers, 1984, p. 281.)

they may be partially or largely filled with a mulch that has been moved by either wind or animals. If the area has a short rainy period and the cracks close, they remain closed for less than 60 consecutive days.

2. *Uderts* (Latin *udus,* humid) are the Vertisols of humid climatic realms. Their cracks open and close at irregular times, depending upon the weather. In fact, some years may experience no cracking at all, although this must occur in less than 50 percent of all years.

3. *Usterts* (Latin *ustus,* burnt) are the Vertisols of tropical and subtropical monsoon climates that experience two distinct rainy and dry seasons, and of temperate regimes that are characterized by limited summer rains. These soils have cracks that open once or twice during the year and remain open for 90 cumulative days or more in most years. The cracks are normally closed for 60 consecutive days or more each year during the wet season.

4. *Xererts* (Greek *xeros,* dry) are the Vertisols of the earth's Mediterranean climates, that is, areas with cool, wet winters and warm to hot, arid summers. They have cracks that close and open regularly once per year in association with the alternating wet and dry seasons. In most summers the cracks will remain open for more than 60 days.

INCEPTISOLS

Inceptisols include soils of widely differing environments in which a variety of pedogenic processes are operative. Some are weathered sufficiently to produce altered horizons that have lost either bases or iron and aluminum, but still retain weatherable minerals. At the same time they

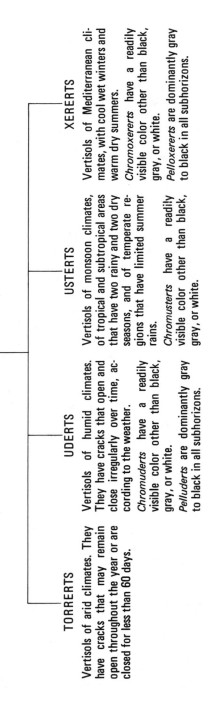

VERTISOLS

These are clayey soils that have deep wide cracks at some time during the year.

TORRERTS

Vertisols of arid climates. They have cracks that may remain open throughout the year or are closed for less than 60 days.

UDERTS

Vertisols of humid climates. They have cracks that open and close irregularly over time, according to the weather.

Chromuderts have a readily visible color other than black, gray, or white.

Pelluderts are dominantly gray to black in all subhorizons.

USTERTS

Vertisols of monsoon climates, of tropical and subtropical areas that have two rainy and two dry seasons, and of temperate regions that have limited summer rains.

Chromusterts have a readily visible color other than black, gray, or white.

Pellusterts are dominantly gray to black in all subhorizons.

XERERTS

Vertisols of Mediterranean climates, with cool wet winters and warm dry summers.

Chromoxererts have a readily visible color other than black, gray, or white.

Pelloxererts are dominantly gray to black in all subhorizons.

Figure 6.7 Suborders and Great Groups of the Soil Order Vertisol.

normally lack illuvial horizons that have been enriched with either silicate clays containing aluminum or amorphous mixtures of aluminum and organic carbon. Inceptisols are generally found in humid climates where leaching is active (Figures 6.8 and 6.9). In short, Inceptisols are beginning to exhibit pedogenic characteristics associated with weathering as compared with Entisols and Vertisols, but these features are too weakly developed to be considered diagnostic of maturely developed soil.

There are six suborders of Inceptisols (Figure 6.10), as follows:

1. *Andepts* are relatively freely drained. They have a low bulk density and are usually formed in pyroclastic materials (such as volcanic ash or pumice), although certain sedimentary and basic extrusive igneous rocks can also serve as the parent material. Andepts are rich in either glass or allophanes, such as amorphous clays. Glass is a unique parent material in that it is relatively soluble in comparison to the chrystalline aluminosilicates, and weathers rapidly to produce amorphous products. Found in any latitude, these soils are restricted to areas in or near mountains with active volcanoes. Repeated ash falls are quite common in Andepts, and these soils may have two or more buried horizons within 1 meter of the surface.

2. *Aquepts* (Latin *aqua,* water) are wet Inceptisols. The natural drainage of these soils is poor or very poor, and the water table usually stands close to the surface at some time during the year. Surface horizons are dominantly gray to black, and subsurface horizons are dominantly gray.

3. *Ochrepts* (Greek *ochros,* pale) are light-colored, brownish, and relatively freely drained Inceptisols of middle to high latitudes. They form in crystalline parent materials and in a moisture regime with an annual excess of precipitation over evapotranspiration.

4. *Plaggepts* (German *plaggen,* sod) include all freely drained soils that contain a plaggen epipedon, except for a few Andepts.

The plaggen epipedon is a man-made surface layer, more than 50 centimeters thick, that has been produced by long-term manuring. In medieval times, sod or other materials was commonly used for bedding livestock, and the manure was spread on the field being cultivated. The mineral materials brought in by this kind of manuring eventually produced an appreciable thickened Ap, as much as 1 meter or more in thickness.

Colors and contents of organic carbon depend on the sources of the materials used for bedding. If the sod was cut from the heath, the plaggen epipedon tends to be black or very dark gray, to be rich in organic matter, and to have a wide carbon-nitrogen ratio. If the sod came from the forested

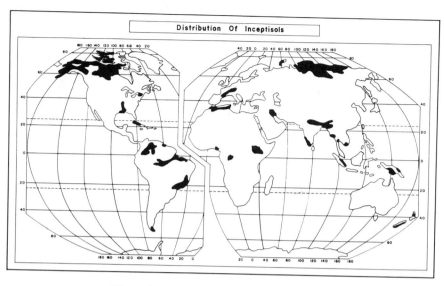

Figure 6.8 World distribution of Inceptisols.

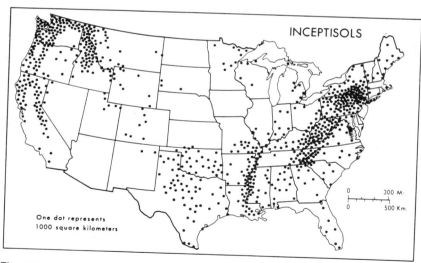

Figure 6.9 Inceptisols of the United States. (From Philip J. Gersmehl, "Soil Taxonomy and Mapping," *Annals of the Association of American Geographers,* 67, September 1977, p. 423. By permission of the Association of American Geographers.)

INCEPTISOLS

These soils have altered horizons that have lost bases or iron and aluminum but retain weatherable minerals and that lack illuvial horizons either enriched with silicate clays that contain aluminum, or those enriched with amorphous mixtures of aluminum and organic carbon.

ANDEPTS

Inceptisols that are relatively freely drained and have appreciable amounts of allophane or pyroclastic materials.

Cryandepts are the cold Andepts of high mountains or high latitudes.

Durandepts have a duripan within 40" of the surface.

Dystrandepts have large amounts of organic carbon and amorphous materials, and a small amount of bases.

Eutrandepts have large amounts of organic carbon and amorphous materials, and an ample supply of bases.

Hydrandepts have perudic moisture regimes.

Placandepts have placic horizon.

Vitrandepts have the largest amounts of vitric ash and pumice and the lowest amounts of their weathering products.

AQUEPTS

These are wet Inceptisols with poor drainage.

Andaquepts are formed in pyroclastic materials.

Cryaquepts are found in arctic and subarctic regions.

Fragiaquepts have a fragipan.

Halaquepts are sodic and saline soils.

Haplaquepts are light colored, gray Aquepts, mostly of humid climates.

Humaquepts are nearly black or peaty, very wet, acid Aquepts.

Placaquepts have a placic horizon.

Plinthaquepts have plinthite present.

Sulfaquepts are acid sulfate soils.

Tropaquepts are found in the tropics.

OCHREPTS

Inceptisols that are freely drained and light in color.

Cryochrepts are found in the cold areas of high mountains or high latitudes.

Durochrepts have a duripan in the upper 40".

Dystrochrepts are brownish, acid soils of humid and perhumid regions.

Eutrochrepts are brownish, base rich soils of humid regions.

Fragiochrepts have a fragipan.

Ustochrepts are found in subhumid to semiarid regions.

Xerochrepts are found in Mediterranean climates.

PLAGGEPTS

Inceptisols that have a plaggen epipedon composed of crystalline rather than pyroclastic materials.

TROPEPTS

Inceptisols that are freely drained, brownish to reddish, and found in intertropical regions.

Dystropepts are acid Tropepts of high rainfall regimes.

Eutropepts have high base saturation and are rarely dry for extended periods.

Humitropepts are found in cool humid regions of high altitude.

Sombritropepts are dark, humus rich Tropepts of perhumid cool, hilly or mountainous regions.

Ustropepts are base rich Tropepts of subhumid climates.

UMBREPTS

Inceptisols which are acid, dark reddish or brownish, freely drained, and organic rich.

Cryumbrepts are found in high latitudes or high altitudes.

Fragiumbrepts have a fragipan.

Haplumbrepts are freely drained and have either a short or no dry season during the summer.

Xerumbrepts are found in Mediterranean climatic regimes.

Figure 6.10 Suborders and Great Groups of the Soil Order Inceptisol.

soils, the plaggen epipedon tends to be brown, to be lower in organic matter, and to have a narrower carbon-nitrogen ratio.

The plaggen epipedon may be identified by several means. It commonly contains artifacts, such as bits of brick and pottery throughout. Chunks of diverse materials, such as black and light gray sand as large as the size held by a spade, may be present. The plaggen epipedon normally shows spade marks throughout as well as remnants of thin stratified beds of sand that were probably produced on the surface by beating rains and later buried by spading. The polypedons that have plaggen epipedons tend to be straight-sided rectangular bodies, and they are usually higher than adjacent polypedons by as much or more than the thickness of the plaggen epipedon (National Cooperative Soil Survey, 1970).

5. *Tropepts* (Greek *tropikos,* of the solstice) are found within the tropics and include freely drained Inceptisols that are brownish to reddish in color. They normally have an ochric epipedon or a cambic horizon and lack significant amounts of active amorphous clays or pyroclastic materials. Within this soil's climatic regime,

> Biologic activity is continuous unless there is a pronounced dry season. Cultivated soils cannot remain bare and exposed to erosion for long periods, for if there is rain the plants start to grow. The continuous biologic activity is reflected in the nature of the organic matter in at least two important ways. One is that for any given kind of soil, C–N ratios tend to be lower in tropical than temperate regions. C–N ratios of virgin soils in the humid tropics compare with those of Aridisols in temperate regions; often they are extremely low. The other difference is that the amount of organic matter is not well reflected by the color of the soil. Black soils may have little organic matter; light-colored soils may be very rich in organic matter, or very poor (Smith, 1965, 41).

6. *Umbrepts* (Latin *umbric,* shade) are acidic, dark reddish or brown, freely drained, organic-rich Inceptisols that occur in humid regions of the middle to high latitudes. These soils have much more organic matter than the Ochrepts, but their base saturation is too low to classify them as Mollisols. They occur in hilly to mountainous regions with relatively high precipitation, even though many experience a distinct dry season during the summer.

LAND UTILIZATION AND MANAGEMENT PROBLEMS

Because they occur in varied sites and climates on a multitude of parent materials, Entisols, Vertisols, and Inceptisols have many uses and pose many diverse management problems. The most significant of these soils, in terms of areal occurrence, are the Vertisols, the Entisol suborder

Aquents, Fluvents, and Psamments, and the Inceptisol suborder Aquepts. Just as water is the primary agent in the formation of the aquic and fluvial suborders (Aquents, Aquepts, and Fluvents), its management is also a major factor in making these soils agriculturally productive. Under natural conditions most have supported a forest cover of water-tolerant trees. When the lands are cleared for agricultural purposes, flooding and poor drainage pose a constant threat to the stability of crop yields and are the chief causes of crop failure unless surface or subsurface drainage systems and flood control measures are effectively employed to remove and regulate surplus water.

Crops grown on these immature soils include cotton, soybeans, corn, oats, wheat, barley, sugarcane, rice, and vegetables. The success of crop production after drainage is largely dependent upon physical conditions of the soil and availability of nutrients for plants.

When utilized for crop production, the primary factors that are detrimental to the soil's physical properties are loss of organic matter, erosion, inadequate drainage, improper management practices, and equipment traffic. In areas of relatively high temperatures and surplus moisture (such as the Mississippi River Delta region), organic materials rapidly decompose when aerated. As a consequence, such soils lose a major portion of their original organic matter shortly after being brought under cultivation. This results in reduced stability of soil structure, decreased water infiltration, and increased surface runoff. The same conditions responsible for depletion of organic material also create difficulties in replenishing or maintaining this soil component.

Although the landscapes of the aquic and fluvial environments normally have gentle slopes, erosion can be a serious problem.

> Most of the organic matter was in the top few inches [5 to 10 centimeters] and has been most affected by erosion. Furthermore, the ratio of sand, silt, and clay in the surface soil provided better physical conditions than the deeper soil. The removal of the surface layer by sheet erosion leaves material with less desirable physical properties exposed or near the surface [Grissom, 1957, 529].

Man's manipulation of these soils with machinery may also lead to their deterioration. If cultivated when too wet, soil structure may break down. In other instances, *pressure pans* may develop from the movement of heavy equipment over wet ground.[4] The resulting compacted layer

---

[4] A *pressure pan* is a subsurface horizon or soil layer having a higher bulk density and a lower total porosity than the soil directly above or below it, as a result of pressure that has been applied by normal tillage operations or by other artificial means. It is frequently referred to as *plow pan, plow sole,* and *traffic pan.*

resists vertical movement of water and root penetration. Consequently, plants may suffer from too much water during rainy periods and from a lack of moisture during interprecipitation periods. One possible solution is *deep tillage,* or subsoiling, which temporarily eliminates the compacted zone. When deep tillage is practiced on compacted soils, the water infiltration rate, field capacity, and root development all increase, and crop yields are generally higher.

Many farmers, in an attempt to improve the physical condition of their soil, have grown winter legumes, which they plow under prior to planting a summer crop. This reduces erosion and increases organic matter and available nitrogen.

As is true with all soils, the availability of nutrients to meet plant demands is an additional factor in the management of the aquic and fluvial lands. The usual primary deficiencies are nitrogen, phosphorus, potassium, and lime. Nitrogen is required for practically all nonleguminous crops and may be supplied either by crop rotations that include legumes or by application of commercial inorganic materials. In general, it is more economical to use commercial nitrogen on cultivated crops.

Psamments may be stabilized sand dunes or soils developed on sandy deposits. In most instances they are of extremely limited agricultural use. Due to the soil's low moisture-holding capacity and low organic content, coupled with a deficient nutrient supply, their potential crop yields are usually low. Psamments' primary use has been for livestock grazing, although in favored climates some soils have been modified by man and now support truck farms and citrus groves.

Vertisols have their maximum occurrence between latitudes of 45° North and South, and are most extensively developed in Australia, India, and the Sudan.

> Agriculturally the soils have great potential where power tools, fertilizers, and irrigation are available. The natural fertility level can be considered quite high, although the use of nitrogen and phosphorus is beneficial. Tillage of the soil is difficult with primitive tillage tools. The "blacklands" of Texas and Alabama are some of the best agricultural lands in the United States. Worldwide, Vertisols are used mainly for cotton, wheat, corn, sorghum, rice, sugarcane, and pasture [Foth and Turk, 1972, 266].

The major problem with using Vertisols relates to their high shrink-swell capacity. The stress created by the alternate expansion and contraction of their clays has been known to break gas and water lines, rupture highway pavement, cave in basements, and displace building foundations.

Aridisols

7

Aridisols are mineral soils with an ochric (pale) epipedon; they contain one of a wide variety of diagnostic subsurface features. Table 7.1 contains definitions of the most common of these pedogenically related forms. The presence of Aridisols is closely associated with the world's climatic deserts (Figures 7.1 and 7.2). Notably, large expanses are found in the Sahara, Namib, Atacama, Sonoran, Australian, Thar, and Gobi deserts. These soils have the shallowest profiles of any soil regionally developed at the order level. Their depth characteristics and pale color are significantly related to regions of high moisture demand, low water supply, and sparse vegetative cover.

CLIMATE AND VEGETATION

Arid climatic regimes occur over approximately 20 percent of the world's land area. These regions have a large disparity between potential evapotranspiration (PE) and environmental water supply. A high incidence of clear skies permits solar radiation to reach land surfaces with minimum depletion. Thus, desert surfaces heat very rapidly after sunrise, raising ground temperatures to high levels and generating very steep, near-surface, atmospheric lapse rates. After sunset, the reverse is true: Clear desert skies permit a rapid loss of energy through terrestrial radiation, and temperatures diminish quickly. As a consequence, temperature patterns can exhibit high ranges and maxima, both diurnally and seasonally.[1]

---

[1] The primary exception is found in coastal deserts where cool ocean currents may modify temperature extremes.

Table 7.1 Significant Features of Soil Classification in the Aridisol Order

Epipedon:
Ochric — Surface horizons that are light in color and low in organic material.

Diagnostic horizons:
Argillic — A subsurface illuvial horizon of the mineral soil in which layer-lattice silicate clays have accumulated by illuviation to a significant extent.

Calcic — Horizons of secondary carbonate enrichment that are > 15 cm thick and have a calcium carbonate equivalent content > 15 percent, and have ⩾ 5 percent calcium carbonate than the underlying C horizon.

Cambic — A mineral soil horizon that has a texture of loamy, very fine sand or finer, has soil structure rather than, rock structure, contains some weatherable minerals, and is characterized by the alteration or removal of mineral material as indicated by mottling or gray colors, stronger chromas or redder hues than in underlying horizons, or the removal of carbonates.

Gypsic — A subsurface horizon that is enriched with calcium sulfate, containing at least 5 percent more gypsum than the C horizon.

Natric — A special kind of argillic horizon that has the properties of (1) prismatic or columnar structure, or occasionally blocky structure; and (2) in some subhorizon has at least 15 percent saturation with exchangeable sodium.

Petrocalcic — A continuous indurated calcic horizon, cemented by carbonates of calcium, and in places with some magnesium carbonate.

Salic — A subsurface horizon at least 15 cm thick, enriched in soluble salts, generally at least 2 to 3 percent depending on thickness.

Other diagnostic features:
Duripan — A subsurface horizon that is cemented by silica, usually opal or microcrystalline forms, to the point that fragments from the air-dry horizon will not slake in water or acid.

Gilgai — Microrelief features of soil produced by expansion and contraction with changes in moisture. Found in soils containing large amounts of clay that shrink and swell considerably with wetting and drying. Usually has the appearance of a succession of microbasins and microknolls.

Lithic contact — A boundary between soil and continuous coherent underlying material with a hardness > 3 Mohs.

Paralithic contact — Similar to lithic contact except that the underlying material is softer (< 3 Mohs).

Slickensides — Shiny ped surfaces created in clay soils by one mass of soil sliding past another.

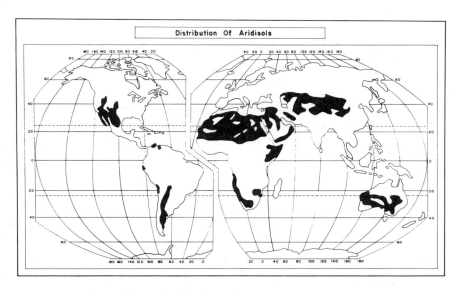

Figure 7.1 World distribution of Aridisols.

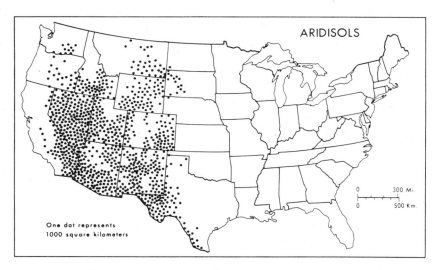

Figure 7.2 Aridisols of the United States. (From Philip J. Gersmehl, "Soil Taxonomy and Mapping," *Annals of the Association of American Geographers,* 67, September 1977, p. 424. By permission of the Association of American Geographers.)

Moisture demand (PE) in deserts is typically high (Figure 7.3), and the amplitude of the moisture demand curve is proportionately much greater than that for temperature. This indicates that temperature change basically behaves in a linear fashion, while the capacity of the atmosphere to evapotranspire and store moisture increases at an increasing rate, that is, exponentially (Figure 7.4). Thus, it can be seen that with temperature maxima usually above the norm for their latitudinal location, the atmosphere of desert regimes possesses high energy availability. The relationship between this energy available for evapotranspiring moisture and the environmental water supply is a critical factor in determining an area's degree of aridity.

Moisture supply in arid regions is erratic, and mean precipitation figures provide but scant information on expected amounts of available water during any given period. In a 10-year span at Las Vegas, Nevada (Table 7.2), for example, annual precipitation ranged from 14.2 millimeters to 141 millimeters. This is a variation from the annual precipitation norm—99.1 millimeters—of 14 and 142 percent, respectively. Individual months experience even greater departures. August has a percentage range from less than 1 to over 539 percent.

The moisture that does reach the surface is not only unreliable and low in total amount, but also generally has minimal opportunity to infiltrate soil and percolate through the profile. Desert landscapes that have been baked by the sun during interprecipitation periods are not often receptive to water infiltration. Furthermore, much precipitation in these areas arrives as thermally induced convective showers of short duration. Hence, surface water either (1) becomes rapidly channelized and infiltrates into the more porous stream beds to recharge ground-water storage, or (2) is lost through evaporation on the hot desert surfaces under conditions of high energy availability. This brief, transient role of surface water in deserts has led numerous researchers to suggest that profile development and mineral alteration of many Aridisols may be largely the result of past climates that were characterized by higher, more uniform rainfall. A reexamination of Figure 7.3 shows that there is little soil moisture available for leaching processes, a factor supporting such thinking.

Stressing the uniqueness of the desert regime, it must be understood that aridity is strictly a phenomenon of climate, not of vegetation or soil formation. It represents deficient moisture and must include the limits of water demand as well as supply. Too frequently, simple measurements of aridity rely solely on an area's precipitation record, which results in an erroneous delimitation of climatic deserts. Such an example would be the inclusion of the Antarctic Plateau, where precipitation (water-equivalents)

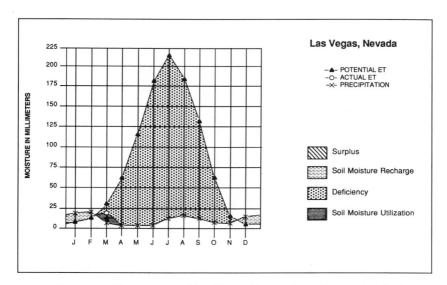

Figure 7.3 Water budget of Las Vegas, Nevada (based on normal data).

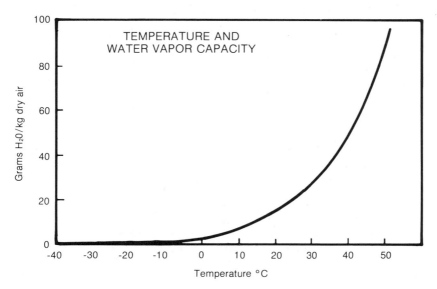

Figure 7.4 Relationship between temperature and water vapor—holding capacity of the atmosphere.

Table 7.2 Precipitation Data for the Period 1951-60 and Precipitation Normal Values for Las Vegas, Nevada

Year	Ja	F	Mr	Ap	My	Je	Jl	Ag	S	O	N	D	Annual
1951	4.8[a]	0.5	0.8	1.0	2.5	0.0	13.7	4.1	24.9	1.3	11.2	5.8	70.6
1952	27.2	T[b]	38.1	14.5	T	0.3	11.9	12.4	22.1	0.0	4.1	10.4	141.0
1953	0.3	0.0	0.8	T	0.8	T	5.8	2.5	0.5	3.0	0.5	T	14.2
1954	23.1	0.5	20.6	T	0.0	0.3	40.9	7.1	11.4	0.5	9.1	6.1	119.6
1955	35.6	3.3	T	2.8	0.8	9.9	39.4	44.2	0.0	T	0.8	0.5	137.3
1956	5.8	0.0	0.0	2.0	T	0.0	41.7	0.0	0.0	2.3	0.0	0.0	51.8
1957	6.6	4.1	2.5	14.0	3.8	T	10.4	65.8	T	12.4	6.6	0.3	126.5
1958	10.9	18.5	14.0	16.3	4.6	0.0	0.3	4.6	5.3	16.0	24.4	0.0	114.9
1959	2.8	18.3	T	T	T	T	0.5	8.4	0.3	13.0	27.7	35.1	106.1
1960	14.2	10.7	1.3	2.5	T	T	10.4	T	5.8	12.4	47.8	6.6	111.7
Normal	13.5	11.2	8.9	5.8	2.0	1.0	12.7	12.2	8.6	5.1	7.9	10.2	99.1

[a] Quantities are given in millimeters.
[b] T = trace of precipitation.

approximates 100 to 125 millimeters, in a desert classification.[2] Obviously, such a designation would be incorrect, even though the area does, in fact, receive very little moisture. One need only question how glacial ice, which exists exclusively in surplus moisture regimes, could accumulate to depths of hundreds of meters if the plateau was, indeed, a moisture-deficient, or arid, region. The obvious answer is that glacial ice could not form unless moisture demands were low and a surplus existed. Thus, while such regions may be deserts in the sense that large portions of them cannot support plant life, they are definitely not climatic deserts.

It is also incorrect to conceive of all deserts as being composed of active, shifting sand dunes. Such sandy ergs make up a minor portion of the world's deserts, less than 25 percent.[3] Practically all of the remaining 75 percent of desert land contains some vegetation. Those plants that do exist in desert environments have a wide variety of adaptive techniques for surviving in moisture-deficient areas. One characteristic that is common to all arid land vegetation, however, is that plants are widely spaced. In general, the greater the intensity of dryness, the more widely separated are individual plants. Thus, there is normally a decrease in total biomass (amount of living matter in a given area) associated with decreased moisture. This factor has important effects upon soil. Its color, structure, microbial population, exchange capacity, and nitrogen content are all heavily influenced by the amount of organic matter that is present.

PEDOGENESIS

Of all the regionally distributed soils, Aridisols exhibit the least depth of horizon and total pedon development. Their formation in an arid climate has meant that certain types of chemical weathering are limited, and excessive leaching of soluble minerals, which would otherwise be removed from the profile, is prevented.

Two groups of soil-forming processes predominate in the arid regime: *calcification* and *salinization.* The first is dominant over extensive segments of the earth's moisture-deficient regions. The second is normally localized within such areas.

---

[2] *Water-equivalent* refers to the area/depth value of snow or ice, when melted.

[3] Unstabilized sand dunes do not exhibit profile development and are not classified in the Aridisol order; rather, they are identified as Entisols. *Erg* is a term referring to sand dune covered deserts of the Sahara.

CALCIFICATION

The processes by which calcium carbonate or calcium and magnesium carbonates may accumulate in the soil profile are called *calcification*. These processes result in the production of either a *calcic horizon* or a *petrocalcic horizon,* the latter known as *caliche* in the southwestern United States. This feature, designated as the ca layer in profile description, is normally found in the C horizon, but it may encroach into the uppermost horizon under intensely arid conditions.

A calcic horizon is a subsurface soil layer of secondary carbonate enrichment at least 15 centimeters thick. The primary genetic factor seems to be limited rainfall that is inadequate to remove lime completely from even the few surface centimeters of the soil. If the parent material is rich in carbonates or is "wind-dusted" with regular carbonate additions, the calcic horizon through time will become plugged with carbonates and becomes cemented into a hard, massive, continuous petrocalcic horizon. Petrocalcic horizons are normally restricted to older landscapes; the more youthful calcic horizons have lime accumulations that are soft and disseminated, or else concentrated in small, hard concretions.

Figure 7.5 offers a model representing current interpretations of caliche formation. Precipitation falling through the atmosphere dissolves

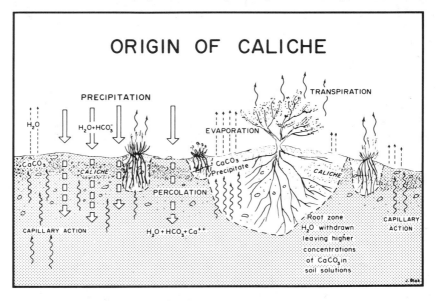

Figure 7.5 A schematic model of Caliche development.

free carbon dioxide, producing a mild carbonic acid. This solution infiltrates unsaturated portions of soil, leaching surface materials and taking ions of the very soluble calcium into solution. The depth to which this solution percolates depends on several variables, including the amount, intensity, and frequency of rainfall; the permeability of the soil complex; and the moisture demands of the environment. Increasing precipitation totals increase the depth at which carbonate accumulation takes place.

With each succeeding precipitation event, calcium carbonate in the accumulation zone increases. Interprecipitation periods, on the other hand, are accompanied by loss of moisture from the solum through evapotranspiration. Consequently, the soil solution becomes concentrated with soluble minerals to the saturation point. Further moisture withdrawals from this saturated solution result in a precipitate of minerals, most frequently calcium carbonates. Subsequent rains continue the process, resulting in a thickening of the carbonate zone in the direction of the land surface. Such accretion ends only when the supply of calcium ions is no longer sufficient to develop the caliche layer further. When this stage of weathering is achieved, the uppermost surface of caliche is repeatedly dissolved and reprecipitated, resulting in a hard crust on the surface of the caliche horizon.

Occasionally very thick deposits of caliche are identified. These are considered to be the product of soils experiencing constant illuviation, hence having available a continuing supply of fresh minerals for the weathering and release of ions. A profile description of a soil with a petrocalcic horizon is shown in Table 7.3.

SALINIZATION

Processes of *salinization* lead to accumulation of mineral salts in the soil profile of sufficient concentration to limit plant production to species with a high salt tolerance, that is, *halophytes.* In arid regions, salinization is closely associated with restricted surface water removal in localized areas of entrapped drainage. It is common for such a condition to exist in *bolsons,* where runoff from precipitation can become imponded in valley bottoms.[4] These enclosed basins, with no existing stream channels, would become permanent lakes in humid regions. In arid climates, water descending from surrounding slopes assimilates soluble minerals and transports them into the basin, where an intermittent lake may be formed

---

[4] A *bolson* is a basin of interior drainage in an arid or semiarid region, whose floor is tending to be filled by a number of *alluvial fans* around its flanks.

Table 7.3   *Profile Description of Soil with Petrocalcic Horizon*

Horizon	Depth (Cm)	Cave Gravelly Sandy Loam (Pima County, Arizona)
A1	0-3	Brown (7.5YR 5/4) gravelly, sandy loam, dark brown (7.5YR 4/4) moist; weak thick platy structure; slightly hard, very friable, nonsticky, nonplastic, common very fine and fine roots; common fine and very fine interstitial and vesicular and few fine tubular pores; 20 percent gravel; strongly effervescent; moderately alkaline (pH 8.0); clear smooth boundary.
C1	3-18	Light brown (7.5YR 6/4) gravelly, fine sandy loam, dark brown (7.5YR 4/4) moist; massive; slightly hard, very friable, slightly sticky, slightly plastic, common very fine, fine, and medium roots; few fine vesicular and interstitial pores; 30 to 35 percent gravel; strongly effervescent; moderately alkaline (pH 8.2); abrupt wavy boundary.
C2cam	18-51	White (10YR 8/2) gravelly, indurated layer with thin laminar layer on upper surface, very pale brown (10YR 7/3) and light yellowish brown (10YR 6/4) moist; massive; extremely hard; very few very fine roots; few fine vesicular pores; violently effervescent; moderately alkaline (pH 8.2); abrupt wavy boundary.
C3cam	51-81	White (10YR 8/2) and very pale brown (10YR 8/3) strongly cemented gravelly layer, very pale brown (10YR 7/3) and light yellowish brown (10YR 6/4) moist; massive; extremely hard to very hard, very firm; very few very fine roots; violently effervescent; moderately alkaline (pH 8.2); clear wavy boundary.
C4cam	81-104	White (10YR 8/2) and very pale brown (10YR 8/3) strongly cemented, gravelly layer, very pale brown (10YR 7/3) and light yellowish brown (10YR 6/4) moist; massive; very hard to extremely hard; very firm; very few very fine roots; violently effervescent; moderately alkaline (pH 8.2).

Source: Hendricks and Havens (1970, 24-25).

during, and immediately after, the precipitation event. During interprecipitation periods, on the other hand, the entrapped water evaporates, leaving behind a residue of mineral salts encrusted on the surface and precipitated within the soil profile. These mineral salts can produce salic or natric horizons. A *salic horizon* must be at least 15 centimeters thick and have a minimum of 2 percent soluble salts. The *natric horizon* is a special type of *argillic* (clay) *horizon.* In addition to the properties of the argillic horizon, it must also have (1) prismatic or columnar structure and (2) in some portion of the horizon a minimum of 15 percent saturation with sodium. Thorne and Seatz (1955) describe the accumulation of mineral salts and their effect on soil as follows:

> The soluble salts that accumulate in soils consist principally of cations of calcium, magnesium, and sodium and the chloride and sulfate anions. Potassium, bicarbonates, and nitrate ions occur in smaller quantities. Borates occasionally occur in small amounts but receive considerable attention because of their exceptionally high toxicity to plants. The proportion of ions occurring in soils varies greatly. The nature of the salts present obviously depends on the composition of the rocks weathered, the nature of the weathering process, and subsequent reactions during the moving of the salts from the site of weathering to the place of deposition.
>
> Salt may influence soils in many ways. The direct presence of salts is one aspect; changes in exchangeable cations on the soil colloids is another; and a third includes the indirect effects of salts on soil microbes, plant root activities, and the physical properties of soil colloids.

Some of the oldest soils in arid regimes show well-formed argillic horizons, while more recent surfaces generally lack such development. This feature perhaps may be a product of past climates with greater rainfall. The *argillic horizon* is an illuvial subsurface horizon in which silicate clays have accumulated. The genesis of this clay-enriched zone in arid climates has been explained by two diverse schools of thought. One point of view is that clays form *in situ* within the B horizon, because this soil layer is deeper, more protected from the sun's evapotranspiration demands, and remains moist for longer periods of time than does the epipedon. These combined factors, it is argued, provide an opportunity for aluminosilicate clay formation from by-products of minerals being altered within a soil layer of active chemical weathering. The other perspective is that subsurface clay enrichment represents a relict soil feature associated with past climates that were humid, and when illuviation processes were more dominant than at present. The basis of disagreement centers on the fact that *clay skins* (cutans) are frequently absent in the Argids (Aridisols with an argillic horizon).[5] Proponents of the first view claim clay skins are

---

[5] *Cutons* are coatings of clay on the surface of soil peds and mineral grains, and in soil pores. They are considered to be evidence of clay migration.

not present because clays formed where they are presently found. The second group's rebuttal is that the argillic horizon doesn't exhibit cutans because they've been destroyed by the repeated shrinking and swelling associated with wetting and drying.

A unique feature of many Aridisols is a residual surface layer of rock and gravel that is produced by deflation of fine particulate matter and is known as *desert pavement.* The remaining exposed rock and stone surfaces commonly have a thin black or dark brown stain from iron and manganese oxides, referred to as *desert varnish.*

Differentiation of Aridisols at the suborder level is based on the degree to which they have been weathered (Figure 7.6). The *Argids* (Latin *argilla,* white clay) are normally found on older surfaces and contain an argillic or natric horizon; these same features are not found in the less developed *Orthids* (Greek *orthos,* true).

Soil associations are groups of defined and named soil taxonomic units occurring together with a characteristic pattern that is repeated throughout a defined geographic region. Figure 7.7 illustrates the soil association concept as it relates to the Beryl-Enterprise Area in Utah. (Note: Similar graphics are found in Chapters 8 through 11.) Soils on this diagram are identified according to their *series name* and relative to their landscape position.[6] The lithosols of mountain uplands are Orthents, that is, Entisols on recent erosional surfaces. The *Preston* and *Richfield* series are, likewise, Entisols. The first is a psamment (Greek *psammos,* sand), a rock-weathered, residual and coarse-textured highland soil; while the latter is a fluvent (Latin *fluvius,* river; English *fluvial,* water-associated) formed on alluvial fans as a product of water-associated erosion and deposition. The remainder of the Beryl-Enterprise Area's soils are classified as Aridisols. The *Sevy* and *Uvada* series are argids, possessing an argillic horizon. Differentiation between them is that the Sevy series has a well-defined B horizon underlain by a petrocalcic layer, whereas the Uvada series of low-lying basins contains a natric horizon with $\geq$ 15 percent saturation of exchangeable sodium. The *Escalante* series are found on toe slopes, are shallow, less mature in development than the Sevy and Uvada series, and have a petrocalcic horizon in close proximity to the surface.

As revealed in Figure 7.7, variation in soil traits is a product of soluble mineral deposition, which is controlled by the time of residence of water delivered to the soil via overland flow.

---

[6] Refer to Chapter 5 and to the Glossary for a definition of *soil* series.

ARIDISOLS

Mineral soils that possess an ochric epipedon and one or more of the following properties: an argillic, calcic, cambic, petrocalcic, gypsic, natric or salic horizon, or a duripan.

ARGIDS

Aridisols that have an argillic or natric horizon.

Durargids have a duripan below an argillic horizon or a natric horizon within 40″ of the surface.

Haplargids have no duripan within 40″ of the surface and have an argillic horizon with less than 35% clay.

Nadurargids have a natric horizon with columnar structure overlying a duripan within 40″ of the surface.

Natrargids have a natric horizon and no petrocalcic horizon reaching with 40″ (1 meter) of the surface.

Paleargids have either a petrocalcic horizon with an upper boundary within 40″ of the surface or have an argillic horizon with more than 35% clay.

ORTHIDS

Aridisols that lack an argillic or natric horizon.

Calciorthids have a calcic or gypsic horizon with an upper boundary within 40″ of the surface.

Durothids have a duripan with its upper boundary within 40″ of the surface.

Paleorthids have a petrocalcic horizon with an upper boundary within 40″ of the surface.

Salorthids have a salic horizon within 30″ of the surface.

Figure 7.6 Suborders and Great Groups of the Soil Order Aridisol.

LAND UTILIZATION AND MANAGEMENT PROBLEMS

The main uses associated with soils of arid climates revolve around either grazing or the intensive production of crops under irrigation in oasis settlements. Grazing predominates over most of the world's Aridisols.

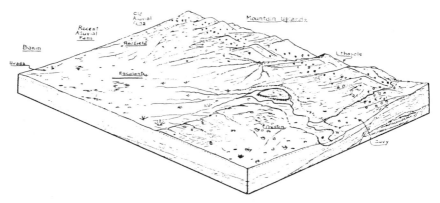

Figure 7.7 Soil Associations—Diagrammatic: Beryl-Enterprise Area, Utah. (Unpublished manuscript of classroom exercises, courtesy of Neil E. Salisbury.)

Grazing operations, however, frequently require some degree of surface irrigation for the production of supplemental feed and for winter grazing. Thus, these two primary land-use functions are not always mutually exclusive of one another.

Grazing activities range from simple nomadic herding in some of the world's developing countries to complex, large-scale commercial ranching in economically advanced areas (Figure 7.8). Aridisols are generally low in organic matter and nitrogen content, the amount reflecting, to a large degree, the intensity of regional moisture deficiency and density of native vegetation. Desert shrub and widely spaced bunch grasses make up the principal vegetative cover for animal consumption. The *carrying capacity* of this vegetation varies greatly.[7] In some areas of the southwestern United States, it takes as much as 40 hectares (100 acres) to supply forage for just one steer. As a consequence, extensive acreage is required to take care of a moderately sized herd.

Animals grazing on desert vegetation can lead to serious deterioration of the biotic and soil resources unless conscientious management practices accompany range use. Overgrazing can speed up erosion and decrease soil permeability. When the plant cover is removed, soils become compacted, and very loose or very wet soils may actually be displaced. This may enhance rapid removal of surface water by overland flow, lead to destructive flooding, and decrease soil moisture storage potential. Management objectives for the protection of range soils have been succinctly summarized by Wasser, Ellison, and Wagner (1957):

---

[7] *Carrying capacity* refers to the number of animals a given area of land can support with its natural vegetation.

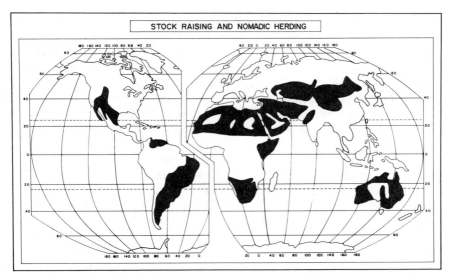

Figure 7.8 World distribution of stock raising and nomadic herding activities.

The basic objective of range management—a dense and well-dispersed cover of vegetation and mulch that provides adequate protection and imparts desirable characteristics to the soil—is economically attainable on most ranges by managing the grazing.

Soil management is accomplished on perhaps as much as 90 percent of the range area almost solely by manipulating grazing animals in accordance with four rules of good range usage: [1] Distributing livestock evenly to insure uniform use of forage on usable portions of each range unit; [2] grazing the kind or kinds of animals that will economically utilize and perpetuate the desirable forage plants growing naturally on each range; [3] adjusting the grazing use of each unit seasonally to meet the growth requirements of the desired forage plants; [4] adjusting the numbers of grazing animals to attain an intensity of forage use that will maintain normal forage and soil productivity.

These considerations are basic to economical livestock production and to the maintenance of forage and soil productivity on all ranges. We call them the cardinal principles of range management.

Aridisols placed under irrigation can be highly productive. Existing in areas with a high incidence of clear skies and high potential photosynthesis, crop yields can be substantially above those in humid climates if an adequate water supply is available.

The manager of arid land must confront problems other than scarcity of water. Saline accumulations in the soil may be at levels that are toxic to plants. The first requirement in reclaiming such land is to install drainage facilities. Subsequent applications of surface water may then remove harmful salts through leaching and subsurface drainage. If the soil has a high percentage of exchangeable sodium, additional treatment with gypsum or sulfur may be necessary. Gypsum provides soluble calcium that replaces sodium absorbed on clay particles. Sulfur applications serve the same basic purpose.

In addition to the foregoing, the farmer may have to level the land, which can expose petrocalcic horizons. The farmer must also deal with nutrient needs of the crops to be to cultivated. In irrigated areas of the southwestern United States, soils are generally nitrogen deficient; most lack sufficient phosphate; susceptible plants commonly experience iron deficiencies or lime-induced chlorosis; and localized zinc, manganese, and boron deficiencies occur.

As if all these management problems, known to every inhabitant of arid regions, were not enough, land managers must also battle the effects of turbulent winds and often heavy downpours.

Mollisols

8

Mollisols are mineral soils that have either a mollic epipedon or its equivalent.[1] A *mollic epipedon* (Latin *mollis,* soft) is a surface mineral layer.[2] It possesses the following properties: (1) The surface retains a soft character; upon drying it never becomes massive and hard; (2) base saturation is high ($\geqslant$ 50 percent); and (3) it is dark colored as a result of abundant humus. These relatively fertile soils are thought to be formed primarily by underground decomposition of roots and surface organic residues (taken underground by animals) in the presence of bivalent cations, particularly calcium.

Most Mollisols are found in intermediate regions between arid and humid climates.[3] As a result of their location, they are transition zones, based on moisture availability that increases in the direction of humid regimes. The area's vegetation reflects this intraregional moisture variation. Although dominated by a surface cover of grasses, the vegetation complex ranges from short, widely spaced bunch grass along arid margins to richer and more luxurient tall-grass prairies near humid boundaries. As might be supposed, the soil exhibits variation in both moisture status and organic matter availability.

---

[1] In certain instances the surface horizon may meet all the specifications for a mollic epipedon except thickness. If this surface horizon lies above an albic layer that is superimposed on an argillic or natric horizon possessing the characteristics of the mollic layer, and if the combined thickness of the mollic layer and the epipedon meet the thickness requirements, the soil is classified as a Mollisol. (See Chapter 9 for a detailed explanation of an albic horizon.)

[2] Although the mollic epipedon is normally a mineral surface horizon, in some very wet soils it may be overlain by an organic layer.

[3] Mollisol boundaries are irregular, with patches and interfingerings into both arid and humid realms in response to unique environmental conditions.

Associated primarily with semiarid and subhumid climatic regions, some of the most extensive contiguous stretches of Mollisols are found in (1) the North American Great Plains and intermontane plateaus, (2) from the western side of the Black Sea eastward beyond Lake Balkhash, (3) Manchuria, and (4) the Argentine Pampa (Figures 8.1 and 8.2).

CLIMATE AND NATIVE VEGETATION

The grassland environment in which Mollisols normally form experiences an annual moisture deficiency and is subject to highly variable precipitation. In certain years rainfall may exceed PE, permitting a temporal expansion of the humid realm and a decrease in the dominance of semiarid/subhumid climates. In other years these patterns may be reversed (Figure 8.3). Precipitation is more reliable than in the arid zone, but still highly erratic. Threat of drought is a continuing reality to grassland farmers, and heavy downpours of rain accompanied by hail and/or tornadoes are not uncommon. An example of such an area's normal moisture supply and demand relationship is illustrated in Figure 8.4.

Avaliable energy for evapotranspiration is largely dictated by latitude and a continental location. Located in the midlatitudes, these areas

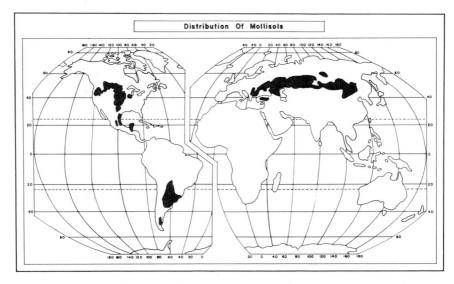

Figure 8.1 World distribution of Mollisols.

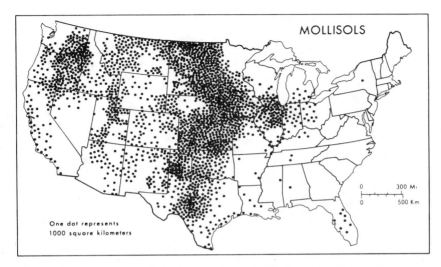

One dot represents
1000 square kilometers

Figure 8.2 Mollisols of the United States. (From Philip J. Gersmehl, "Soil Taxonomy and Mapping," *Annals of the Association of American Geographers,* 67, September 1977, p. 425. By permission of the Association of American Geographers.)

experience pronounced seasonal contrasts in solar heating. In addition, the low specific heat traits of continental interiors permit atmospheric temperatures to respond rapidly to radiation variation.[4] In many ways, the thermal properties of grassland environments parallel those of arid climates, except perhaps for absolute temperature maxima.

A grassland plant community in association with Mollisols does not bear a simple relationship to mean annual precipitation. Rather, soil moisture distribution within the pedon seems to be the crucial factor governing the presence of grasses. When the upper layers of the soil are moist during a considerable part of the year, yet deeper layers are too dry for deep-rooted trees, a grassland can form and be dominant. Porous soils that allow moisture to percolate deeply may, on the other hand, give support to trees in a region that is normally grass covered. Thus, the distribution of grasses is complex. They will, in general, withstand greater environmental stress than forests— not only more extreme aridity, but also more intense cold—and they will grow in wetter places.

---

[4] *Specific heat* refers to the amount of energy required to produce a given temperature increase in a specific substance; for example, it takes approximately 580 calories per cubic centimeter to raise the temperature of water 1 °C. It is estimated that the specific heat of water is aboout five times as great as that for the materials of the continents.

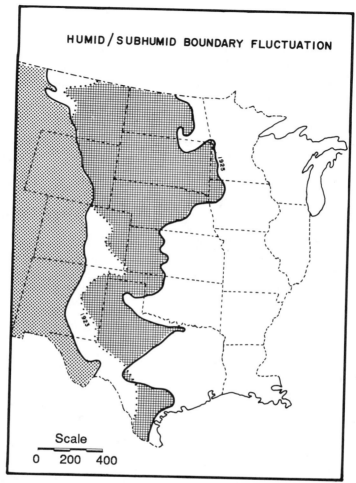

Figure 8.3 Fluctuations in the humid/subhumid boundary in the Great Plains of the United States.

The character of the Mollisol pedon is determined as much by the vegetation as it is by climate. The softness of the epipedon, its coloration, and its available nutrients are all intimately related to the rooting and feeding traits and the life cycle of grasses. The multitude of variation in grassland plants brings about a wide variety of soils. In North America alone a minimum of seven grassland associations have been recognized. However, there are two major grassland types—the true (tall-grass) prairie and the mixed (short-grass) prairie, often called a *steppe*.

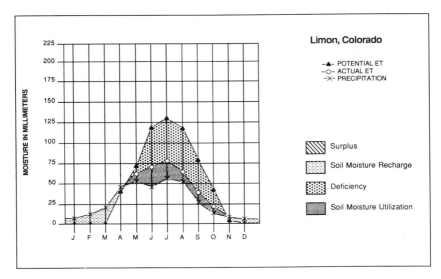

Figure 8.4 Water budget of Limon, Colorado (based on normal data).

Tall-grass prairies exist in subhumid and humid lands. They are found primarily where soil moisture is deeper than in short-grass lands but not deep enough to support trees. Their areal extent is most well expressed on the drier margins of midlatitude forests. Smaller units exist within the boundries of humid areas where calcareous parent material is present or where the ground-water table is exceptionally high. The latter accounts for only a small percentage of the total area in which such vegetation occurs. Approximately 42 percent of all Mollisols have a tall-grass prairie cover; the remainder is primarily short-grass (steppe) vegetation, with forest inclusions.

Short-grass vegetation occurs in semiarid areas, normally adjacent to deserts, where rainfall is sufficient only to moisten the upper soil layers during the warm season and to provide just enough support for shallow-rooted short grasses to survive. As aridity increases, luxuriousness of grasses decreases, and individual plants tend to be more widely spaced and of lower stature than their counterparts in more humid areas.

PEDOGENESIS

The most striking Mollisol feature, and the singular item of taxonomic importance at the order level, is their dark-colored epipedon. This is a product of the unique combination of the rooting habits, nutrient storing capacity, and the growth and decay cycles of grass coupled with the area's moisture regime.

Grasslands differ from forests in both total amount of organic matter incorporated within the solum and in its distribution throughout the pedon. In forested areas, the major contributing source of soil humus is the leaf fall that accumulates on the surface. Percolation of rainfall through this vegetative mass, or its mechanical mixing by soil fauna, results in organic detritus being incorporated in the profile. Grasses also produce an organic mat as a result of the accumulation of their decaying aerial parts. In contrast to quasipermanent and widely distributed rooting structure of trees, however, the dense, fibrous masses of grass roots prolifically permeate and are uniformly concentrated throughout the profile. As these plants die and decay in an almost continuous process, large quantities of humus are produced. It is estimated that even long-lived grasses (such as Bluestem) annually contribute between 600 and 1,000 kilograms of raw organic matter per hectare to the profile, and in Wisconsin, under tall-grass prairie, totals of 136 metric tons per hectare have been recorded. Although most grass roots and organic matter accumulate in the upper 60 centimeters of the pedon, some grasses (such as *Andropogan gerardi*) extend to depths exceeding 200 centimeters. In certain areas, the profile may also exhibit a subsoil of secondary organic matter enrichment, a result of both eluviation of the E horizon and root extension and proliferation during periods of drought (when the A horizon is depleted of moisture).

Even though total amounts of organic material in grassland soils are relatively high, actual content in the epipedon varies considerably. Along arid borderlands where moisture availability is low and vegetation widely spaced and short in stature, the minimum of 1 percent is approached, whereas totals of 8 to 10 percent are common in moister regions, with even higher amounts in truly humid climates. Degree of *melanization*—darkening of the epipedon by organic matter—is reflected by darker hues of brown and an increased nitrogen content as soil organic matter increases. For example, under virgin conditions in the forested southeastern United States, soils average about 1,814 kilograms of nitrogen per hectare, whereas under a Boroll (Chernozem), amounts of 7,250 kilograms per hectare have been recorded.

Grasses, in general, require greater quantities of basic mineral nutrients—particularly calcium—than do trees. As a result, their organic remains are richer in base nutrients, which upon mineralization are subsequently returned to the soil. These related bases are again available for reuse by plants in a continuing cycle that under natural conditions maintains a relatively high degree of base saturation.

The structure of the mollic epipedon is granular (crumb) in those soils that have a texture coarser than silty clay loam, and is fine subangular blocky in textures that are finer. The processes involved in structure formation have been described as follows:

Wet-dry, shrink-swell, freeze-thaw and root expansion—root decomposition cycles in clayey A1 horizons foster the blocky structure. Prairie vegetation pumps available water out of a soil to a depth of several feet (a meter or two). Granular structure is formed by the passage of soil material through earthworms; by microbial production of soil-binding gums and other materials from biotic tissues; by formation of sesquioxide colloidal masses through weathering of primary minerals and mineralization of organic matter; by coagulation of humates by carbonates; by formation of binding organo-clay complexes; by knitting together of soil particles and aggregates by interlacing roots, some of which pierce the peds [Hole and Nielson, 1967, 29].

Character of the Mollisol's eluviation-illuviation products can be explained largely in terms of the "materials being weathered" and the "weathering environment." Several writers have stressed that grassland soils do not normally form from hard rocks such as granite, but rather from unconsolidated calcareous sediments, with loess being the most extensive initial material worldwide. Under tall-grass prairie, leaching of the soil column eventually results in removal of carbonates and bases relative to their original content. Consequently, the soil column loses mass and experiences a pH reduction. In the more extensive dry areas where potential evapotranspiration well exceeds precipitation, moisture rarely percolates below the root zone. Only infrequent wet periods leach the very soluble salts of sodium and potassium from the profile. The less-soluble calcium salts, in particular calcium carbonate ($CaCO_3$), have a tendency to accumulate in lower horizons because normal rainfall reaches only a few decimeters into the subsoil, where it gradually evaporates into the soil atmosphere or is utilized by plants. Under such dehydrating conditions, calcium carbonate is carried into the subsoil in solution and later precipitated into hard nodules as the soil solution becomes supersaturated with mineral concentrates. A continuation of this process may eventually lead to development of a petrocalcic horizon, similar to those produced in many Aridisols. It is evident that the drier Mollisols share pedogenic processes parallel to those dominating the Aridisol regime. A description of a Mollisol of the Typic subgroup of Paleustolls, and located in Cottle County, Texas follows.

As previously stated, the basic classification requirement for all Mollisols is the presence of a mollic epipedon. This criterion alone,

Table 8.1   Profile Description of a Mollisol

Horizon	Depth (Cm)	Description
Ap	0-15	Brown (7.5YR 4/4) silty clay loam, dark brown (7.5YR 3/3) moist and crushed; weak medium subangular blocky structure; very hard, firm; common roots; few strongly cemented $CaCO_3$ concretions up to 6 mm in diameter; few siliceous pebbles and cobble-stones on the surface and in the soil; mainly noncalcareous in matrix but weakly effervescent surrounding $CaCO_3$ concretions; abrupt smooth boundary.
B21t	15-36	Dark reddish brown (5YR 3/2) silty clay, (5YR 3/3) moist and crushed; weak, coarse, prismatic structure parting to moderate medium and fine angular blocky structure; very hard, very firm; few roots; few strongly cemented $CaCO_3$ concretions as much as 6 mm in diameter; few siliceous pebbles; cracks up to 3 cm wide extend through lower boundary; noncalcareous in matrix; clear smooth boundry.
B22t	36-61	Reddish brown (5YR 4/4) silty clay, dark reddish brown (5YR 3/4) moist; strong, coarse prismatic structure parting to strong medium and fine angular blocky structure; extremely hard, very firm; few fine roots; few strongly and weakly cemented $CaCO_3$ concretions; few siliceous pebbles; cracks up to 2cm wide extend through lower boundary.
B23t	61-96	Reddish brown (5YR 4/4) clay, dark reddish brown (5YR 3/4) moist; strong coarse prismatic structure parting to strong coarse and medium angular blocky structure; extremely hard, very firm; few fine roots, mainly between peds; few strongly and weakly cemented $CaCO_3$ concretions; few siliceous pebbles; cracks up to 2 cm wide; calcareous in matrix; gradual smooth boundary.
B24t	96-127	Red (2.5YR 4/6) clay, dark red (2.5YR 3/6) moist on ped faces; streaks of dark reddish brown on faces of peds; interior of peds are red (2.5YR 4/6) moist; moderate coarse prismatic structure parting to moderate coarse and medium angular blocky structure; extremely hard, very firm; few fine roots; few strongly and weakly cemented $CaCO_3$ concretions; few siliceous pebbles; cracks up to 2 cm wide extend to lower boundary; calcareous in matrix; clear wavy boundary.

Table 8.1 (Continued)

Horizon	Depth (Cm)	Description
B25tca	127-152	Red (2.5 YR5/6) clay, red (2.5YR 4/6) moist, red (5YR 4/6) $CaCO_3$ coatings and dark reddish brown (5YR 3/2) clay coatings on faces of peds; moderate medium and coarse blocky structure; extremely hard, very firm; stringers or chimneys of yellowish-red (5 YR 5/6) moist $CaCO_3$ up to 15 cm apart and 1 cm in diameter; cracks extend to the tops of stringers of $CaCO_3$; estimated 5 percent of visable powdery $CaCO_3$ and few strongly cemented concretions; few dark pellets; few siliceous pebbles; calcareous in matrix; gradual wavy boundary.
B26tca	152-178	Dark red (2.5Y 3/6) gravelly clay, light red (2.5Y 6/6) coatings 1 to 2 mm thick of $CaCO_3$; moderate coarse angular blocky structure; very hard, very firm; few fine roots, more piping of $CaCO_3$ in form of powdery (5YR 4/6) bodies than in horizon above; about 15 percent of strongly cemented $CaCO_3$ concretions make up most of the gravel; few siliceous pebbles; few calcareous cobblestones; few dark pellets; few gysum crystals; calcareous in matrix; abrupt wavy boundary.
B27tca	178-198	Dark reddish brown (2.5YR 3/4) clay, black discontinuous coatings on faces of peds; moderate medium angular blocky structure; extremely hard, very firm; a layer high in coarse fragments contains carbonate pebbles up to 8 cm in diameter, 15 percent siliceous pebbles, and a few cobblestones of both carbonate and siliceous rocks; few gypsum crystals; calcareous in matrix; abrupt wavy boundary.
IIB3a	198-208	Dark reddish brown (2.5Y 3/4) clay, yellowish red (5YR 4/6) moist; calcareous coatings 1 to 2 mm thick; very fine and fine mottles of olive gray; moderate medium subangular blocky structure; extremely hard, very firm; few fine roots; few gypsum crystals; abrupt wavy boundary; few tongues extend into IIC horizon.
IIC	208-229	Variegated grayish green (5GY 5/1) moist and reddish brown (2.5YR 3/4) moist; light clay; reddish yellow stains and thin seams of $CaCO_3$; clay coats on some faces; weak platy or blocky structure, retains part of apparent original rock structure; few gypsum crystals; noncalcareous in matrix.

however, is not sufficient. Certain soils with mollic epipedons may be classified in another order if they have other diagnostic features of greater taxonomic significance—perhaps an oxic horizon. (See Chapter 11 for an explanation of an oxic horizon.) There are seven recognized suborders of Mollisols (Figure 8.5): the *Albolls* (Latin *albus,* white) are normally associated with Spodosols and possess an albic horizon (see Chapter 9); *Aquolls* (Latin *aqua,* water) are found in areas of poor drainage and have characteristics related to wetness; the cold grasslands foster development of *Borolls* (Greek *boreas,* northern); *Udolls* (Latin *udus,* humid), *Ustolls* (Latin *ustus,* burnt), and *Xerolls* (Greek *xeros,* dry), are distinguished from one another on the basis of soil moisture. Udolls have the least number of consecutive dry days, Xerolls always have over 60 consecutive dry days annually, and Ustolls are intermediate in dryness. The last suborder *Rendolls* (Rendzina) does not exhibit characteristics typical of long-term weathering; rather the solum is developed from relatively recently exposed calcareous materials.

Figure 8.6 is illustrative of a Mollisol landscape intermixed with soils representing four additional soil orders. It is not uncommon for a multiple representation of orders to exist within a limited geographic area. Indeed, diversity rather than uniformity characterizes soil distribution. Even so, Mollisols are dominant and account for 70.1 percent of the *soilscape,* that is, the soil landscape.

Clarion and *Tama* series are *Udolls*; the former is a *Hapludoll,* the latter an *Argiudoll.*[5] *Clarion* forms in calcareous glacial till of a recent drift plain and contains subsurface calcareous concretions, sometimes referred to as *soft caliche.* Being of the great group Hapludoll, they possess a brownish cambic horizon. *Tama* soils occur in irregular flat upland tracts of loess and are associated with well-drained sloping lands. Their great group designation (Argiudoll) identifies the presence of a well-developed argillic horizon.

Both soils have a dark, granular, friable, mollic epipedon as a result of their formation under grass. *Tama* are older and exhibit a thicker A horizon. Although both have upper layers that evidence leaching having taken place, the grasses (under natural vegetation) or the grain crops have the propensity to bring bases to the surface, regenerating the soil fertility. Organic colloids are abundant in the A horizon, adding to the fertility and resource value. Some of these colloids appear to have been eluviated to the B horizon. Clay minerals have also been eluviated.

---

[5] Udolls are Mollisols that are dry for as much as 90 cumulative or 60 consecutive days per year.

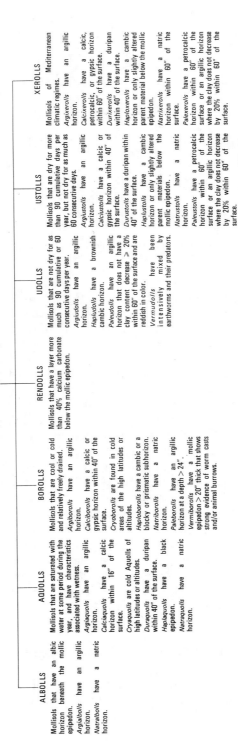

Figure 8.5 Suborders and Great Groups of the Soil Order Mollisol.

MOLLISOLS

Mineral soils that lack an oxic horizon and that have a mollic epipedon or its equivalent.

ALBOLLS

Mollisols that have an albic horizon beneath the mollic epipedon.

Argialbolls have an argillic horizon.

Natralbolls have a natric horizon.

AQUOLLS

Mollisols that are saturated with water at some period during the year, and have characteristics associated with wetness.

Argiaquolls have an argillic horizon.

Calciaquolls have a calcic horizon within 16" of the surface.

Cryaquolls are cold Aquolls of high latitudes or altitudes.

Duraquolls have a duripan within 40" of the surface.

Haplaquolls have a black epipedon.

Natraquolls have a natric horizon.

BOROLLS

Mollisols that are cool or cold and relatively freely drained.

Argiborolls have an argillic horizon.

Calciborolls have a calcic or gypsic horizon within 40" of the surface.

Cryoborolls are found in cold areas of the high latitudes or altitudes.

Haploborolls have a cambic or a blocky or prismatic subhorizon.

Natriborolls have a natric horizon.

Paleborolls have an argillic horizon at a depth > 24".

Vermiborolls have a mollic epipedon > 20" thick that shows strong evidence of worm casts and/or animal burrows.

RENDOLLS

Mollisols that have a layer more than 40% calcium carbonate below the mollic epipedon.

UDOLLS

Mollisols that are not dry for as much as 90 cumulative days per year.

Argiudolls have an argillic horizon.

Hapludolls have a brownish cambic horizon.

Paleudolls have an argillic horizon that does not have a clay content decrease ≥ 20% within 60" of the surface and are reddish in color.

Vermudolls have been intensively mixed by earthworms and their predators.

USTOLLS

Mollisols that are dry for more than 90 cumulative days per year, but not dry for as much as 60 consecutive days.

Argiustolls have an argillic horizon.

Calciustolls have a calcic or gypsic horizon within 40" of the surface.

Durustolls have a duripan within 40" of the surface.

Haplustolls have a cambic horizon or only slightly altered parent materials below the mollic epipedon.

Natrustolls have a natric horizon.

Paleustolls have a petrocalcic horizon within 60" of the surface or an argillic horizon where the clay does not decrease by 20% within 60" of the surface.

Vermustolls have > 50% of the epipedon and > 25% of the underlying horizon composed of worm casts or filled animal burrows.

XEROLLS

Mollisols of Mediterranean climatic regimes.

Argixerolls have an argillic horizon.

Calcixerolls have a calcic, petrocalcic, or gypsic horizon within 60" of the surface.

Durixerolls have a duripan within 40" of the surface.

Haploxerolls have a cambic horizon or only slightly altered parent material below the mollic epipedon.

Natrixerolls have a natric horizon within 60" of the surface.

Palexerolls have a petrocalcic horizon within 60" of the surface or an argillic horizon where the clay does not decrease by 20% within 60" of the surface.

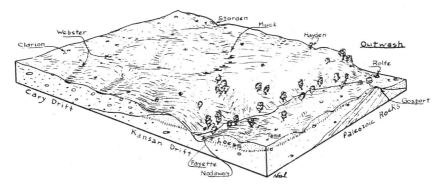

Figure 8.6 Soil Associations—Diagrammatic of Polk County, Iowa (Unpublished manuscript of classroom exercises, courtesy of Neil E. Salisbury).

The *Webster* series are Aquolls; that is, they are saturated with water at some period during the year and have characteristics associated with wetness. They formed in poorly drained swales of the new till plain where numerous sites were provided for the accumulation of ponded waters, and of rank vegetative growth. Because of the location of these sites on the prairie, the vegetation consists mainly of grasses and sedges and other small herbaceous plants.

Organic accumulations in the epipedon are probably the greatest of any nonbog type. These soils possess the darkest and thickest A1 horizons that may be found, and the fertility is perhaps the highest under virgin conditions of any soil in the world.

The *Rolfe* series (*Argialbolls*) is of relatively minor importance, occupying only 0.1 percent of Polk County's area. It is old, hydromorphic soil with a well-developed argillic horizon, as well as an albic horizon that has formed directly beneath its mollic epipedon.

Characteristics of the three other soil orders present in this area are discussed in remaining chapters. At this point, they will be identified only by series name and great group nomenclature. The *Nodaway* and *Storden* series are Entisols (*Udifluent* and *Udorthent,* respectively); *Muck* is a Histosol; *Gosport* is an Inceptisol (*Dystrochrept*); and *Fayette* and *Harden* are both Alfisols (Hapludalfs).

LAND UTILIZATION AND MANAGEMENT PROBLEMS

The high base saturation of Mollisols and the semiarid-to-subhumid climate conditions that encourage growth of a natural cover of grass have been conducive to large-scale commercial grain farming and livestock grazing ventures (especially along the desert margins) or a combination of the two.[6]

Interestingly, these relatively fertile soils were once avoided by farmers in many parts of the world in preference for forested land. A major reason was the difficulty in turning over grassland topsoil. The wooden plow, in use at the time, was incapable of cutting the dense sod mat of grass roots. Consequently, it was easier for the farmer to fell trees and clear forest land for crops. With the invention of the steel plow, slicing and turning over sod were no longer difficult. The steel plow, combined with increased demands for foodstuffs and greater efficiency in transporting agricultural commodities to market, fostered the rapid agricultural settlement in most of the world's midlatitude grasslands in the latter half of the nineteenth century.

Grains grown extensively on Mollisols include wheat, rye, barley, corn, and sorghum, each having areal limitations based on unique habitat requirements. Wheat needs a cool, moist period during early stages of growth, and a hot, dry harvest season. For the seed to ripen, ninety frost-free days and an average summer temperature exceeding 14°C (57°F) are required. The moisture demands of the plant vary with the thermal regime. On poleward margins, greatest wheat growth occurs when precipitation is between 250 and 1,000 millimeters. This is the region of spring wheat, planted in the spring after the last frost and harvested in late summer. Most of the world's wheat, however, is of the winter variety. It is planted in the fall, begins its growth prior to winter, and is harvested early the following summer. Precipitation limits for winter wheat generally range between 500 to 1,750 millimeters (20 to 70 inches).

Even though wheat grows well in semiarid lands, the greatest wheat yields are actually produced in moist, marine west-coast climates. The much desired hard wheats with their high protein content, however, do not thrive in these wetter climates. The greatest advantages of dry grasslands for wheat farming are the extensive expanses of level land suitable for use of large machinery and the area's unsuitability to moisture-demanding crops, such as corn.

Corn is a dominant grain on grasslands where precipitation is either above or very near evapotranspiration demands and where the average summer temperatures are above 19°C (66°F). Of tropical origin, corn grows best under summer conditions that are hot and humid. Rye and

---

[6] In large expanses of the grasslands there is sufficient evidence to indicate that early humans, in controlling vegetation through the use of fire, modified the natural vegetation from a forest cover to one of grass.

barley are normally concentrated on poleward margins of the grasslands, because they tolerate lower temperatures. Grain sorghum is an important crop on Mollisols, especially in the southwestern United States. These drought-resistant grains were introduced into the country from semiarid parts of Africa. They are used for fodder and as a binder crop in strip and terrace cultivation. In dry years grain sorghum provides a partial crop for feed and cover even when wheat may wither and die. It has been known to grow on as little as 250 millimeters (10 inches) of annual rainfall.

In the United States ranching is an important function on Mollisols. Much land is grazed (by cattle, sheep, and goats) on the native grass vegetation. In addition, many land operations have a significant number of hectares committed to seeded and irrigated pastures. Where water supply is dependable, frequent but discontinuous patches of irrigated alfalfa are located adjacent to major streams. Land distant from available surface water usually needs pumped ground water to support sprinkler irrigation systems for pastures. Cash crops grown under irrigation include sugar beets, flax, cotton, peanuts, grain sorghum, melons, onions, and other vegetables. Mollisols in the semiarid southwestern United States also support pecan groves.

The average size of an operating unit for commercial grain farming in the semiarid regime is generally larger than its counterpart in any other area. Farms range from over 400 hectares in the United States to well over 1,200 hectares in the Soviet Union.

Management problems for the semiarid land agriculturist include unreliability of moisture resources, as well as having to maintain favorable fertility and structure characteristics typical of the native Mollisol. Farmers who occupy these areas have had to develop techniques distinctly different from those of the humid lands from which many had emigrated. They soon found that the region's meager precipitation would not permit continuous cultivation of crops, and they adjusted to a land-use pattern called *dry farming.* This technique allows the land to lie fallow for a year, permitting moisture to accumulate and to be stored for production of a crop the following year. The cropland left bare is carefully cleared of vegetation, plowed, and harrowed in such a way to create a dust mulch on the surface, thereby reducing evaporation.

The most common problem facing the agriculturist of the grasslands, especially along its arid borders, is the continual threat of drought. Periods of deficient moisture frequently extend for several months, and occasionally several years may pass without any significant rainfall.

Spodosols

9

Spodosols are mineral soils having either (1) a *spodic horizon* or (2) a *placic horizon* cemented by iron, overlying a *fragipan,* and meeting all the requirements of a spodic horizon except thickness.[1]

A spodic horizon is generally located beneath an eluvial mineral layer, which under virgin conditions is frequently a light-colored *albic horizon,*[2] and is a soil stratum in which "active" *amorphous* (without crystalline structure) materials composed of organic matter and aluminum, with or without iron, have precipitated.[3] A placic horizon, on the other hand, is a thin black to dark reddish pan that is cemented primarily by iron.[4] This feature has a pronounced wavy to involuted form, generally ranges from about 2.5 to 10 centimeters in thickness, and is normally found within the upper 50 centimeters of mineral soils.

The spodic and placic horizons are absent in soils of arid regions, yet are found in humid areas ranging from the tropics to the tundra margins. They are best developed and exhibit maximum areal extent in humid regimes that are both cold and forested; consequently they are dominant in

---

[1] A *fragipan* (modified from *Latin fragilis,* brittle, and *pan,* meaning brittle-pan) is a subsurface horizon that is low in organic matter, high in bulk density relative to the horizons above, and seemingly cemented when dry, having hard or very hard consistency. Their genesis is obscure, but their hardness is largely attributed to the close packing and binding by clays, and their brittleness is thought to be due to weak cementation.

[2] An *albic horizon* is the one from which clay and free iron oxides have been removed, or in which the oxides have been segregated to the extent that the color of the horizon is determined by the color of the primary sand and silt particles rather than by coatings on these particles.

[3] The Soil Survey staff uses the term *active* in this situation to describe materials having a high exchange capacity, large surface area, and high water retention.

[4] A *pan* is a layer or horizon within a soil that is firmly compacted and/or is very rich in clay.

the northern reaches of North America and Eurasia (Figures 9.1 and 9.2). Due to the absence of extensive land masses in the middle latitudes of the Southern Hemisphere, no parallel extensive spodosol occurrence is found south of the equator.

CLIMATE AND NATIVE VEGETATION

The most extensive stretches of Spodosols lie within subarctic climatic regions. It is here that winters are long and cold, there is great seasonal variation in air temperature, and precipitation is concentrated in the summer. The subarctic experiences earth's greatest temperature ranges. Verkhoyansk in the Soviet Union, for example, has a mean monthly temperature variation from approximately −49°C (−57°F) in January to about 14°C (58°F) in July—a range of 63°C (115°F). A similar, although not as intense, pattern is illustrated in the water budget of Figure 9.3. Summers are short, with normally less than four months having temperatures above 10°C (50°F). Warm intervals may raise daily temperature maxima above 48°C (80°F), but at the same time frost may occur on any given day. Winter temperatures are very low, being lower over only permanently ice-covered areas such as are found in Antarctica and Greenland.

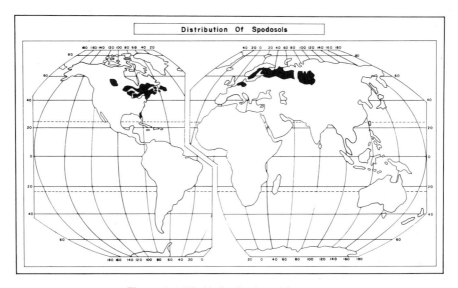

Figure 9.1 World distribution of Spodosols.

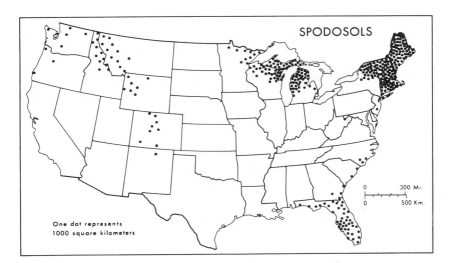

Figure 9.2 Spodosols of the United States. (From Philip J. Gersmehl, "Soil Taxonomy and Mapping," *Annals of the Association of American Geographers,* 67, September 1977, p. 425. By permission of the Association of American Geographers.)

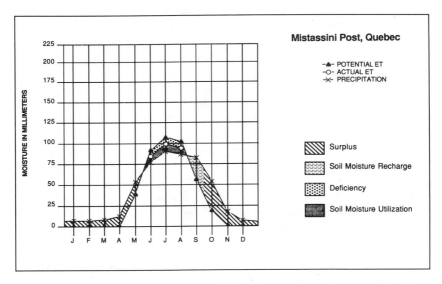

Figure 9.3 Water budget of Mistassini Post, Quebec (based on normal data).

Subarctic precipitation is relatively light. Low temperatures, however, result in minimal evapotranspiration demands, so the surface, amply supplied with moisture, experiences an annual water surplus.

Evergreens dominate in the subarctic region and constitute the most extensive forest formation (often called the *boreal forest*) on the earth's surface. These plants, with their xeric (low moisture)-adapted needle leaves, successfully outcompete broadleaved trees in this environment, where water is unavailable during the part of the year that soils are frozen. Evergreens (as the name implies, they do not shed their leaves at one time) have an advantage over deciduous trees, which shed their leaves during winter. Leaf-shedding trees must grow new sets of leaves before they can resume photosynthesis; thus they do not have maximum use of the short growing season.[5]

The boreal forest zone has two distinct plant formations—one in North America and the other in Eurasia. In the eastern boreal forest of North America, white spruce and balsam fir occupy well-drained sites, while black spruce, jack pine, larch, and tamarack dominate the areas of poor drainage. The Eurasian formation's composition is fairly simple in western Europe, where pine and spruce dominate the landscape. Farther east other species begin to take over, particularly the Siberian fir, spruce, larch, and the dwarf Siberian pine.

The uniqueness of the boreal forest is uniformity. Throughout extensive stretches of land, dense homogeneous stands of evergreen *conifers* (cone-bearers, as most evergreens are) overshadow the soil surface and allow little sunlight to penetrate their canopy to support undergrowth. Therefore, only small herbaceous plants and fungi that require minimal direct sunlight can thrive. Under these shaded conditions and relatively low temperatures throughout a significant portion of the year, microbial activity is low and surface organic debris decomposes slowly. Thus, the mineral soil is blanketed with litter comprised of needle leaves and semidecomposed branches. Decomposition is sufficiently limited that annual leaf falls contribute to an ever deepening mat of organic matter.

Shade provided by forest canopies shields the ground from direct sunlight. Evaporation is thus inhibited, and soil is normally in a moist to wet state, thereby encouraging leaching of bases from the pedon. Most coniferous trees can survive on soils of relatively low base status. Frequently they are anchored within sites consisting of highly siliceous

---

[5] In certain poleward areas winters are so severe that leaves are a disadvantage to plants, and only slower growing deciduous species, such as birch, aspen, and larch, are capable of surviving.

materials. This, of course, means that their organic products (leaf litter, etc.) accumulating on the surface are low in bases and will produce acidic humus upon decomposition. Spodosols do not form *only* under coniferous vegetation. In the subarctic, soils beneath heath also show features typical of this soil order, and in warm climates spodic horizons may exist under savanna, palms, and mixed forest covers. The presence of a spodic horizon in some tropical soils has generated controversy over its pedogenesis in warm areas. Some researchers consider the tropical spodic horizon a product of a past—and different—bioclimatic environment.

PEDOGENESIS

Under optimal conditions a spodic horizon may form in as little as a few hundred years, indicating that pedogenic processes are highly active. The processes influencing development of Spodosols in cool, moist climates are known collectively as *podzolization* (Russian *Podzol,* ash). This may seem a strange way to describe a soil or soil-forming processes, unless one remembers that the original use of the term by pedologists was to describe the soil's color—ashen gray. In nineteenth-century Russia, the soils first classified as podzol had surface mineral horizons from which most minerals, except silica, had been removed via leaching. The remnant, residual product was a sandy, bleached A horizon. Physical mixing of organic matter into the soil stratum (e.g., by burrowing animals or percolating water) subsequently imbued the horizon with a grayish color. Today, this light-colored layer is called the *albic horizon.* Although an albic horizon is frequently found within Spodosols, its presence is not an absolute prerequisite for classifying a soil within this order. Necessary, however, to produce the typical Spodosol morphology are processes involved in the movement and accumulation of sesquioxides and organic matter.

Prior to humus and sesquioxides changing soil location, exchangeable bases must first be displaced by hydrogen ions (H^+) and leached from the upper pedon. The ideal setting for such processes is where there is repeated wetting of the solum (a flushing moisture regime), where temperatures are sufficiently low to inhibit microbial decomposition of organic matter, and where flora has limited demand for base nutrients. Subarctic regions best fulfill these conditions. They have vegetation of coniferous forest with moss, lichens, and subshrubs, and where underbrush and grass are few in number or completely absent. Organic litter produced by these plants is calcium and nitrogen deficient and does not readily decompose. It has compounds of lignin, wax, and resin, which are resistant to breakdown, and may contain tannins and terpenes, which further inhibit decay. The nature of the subarctic boreal vegetation retards biochemical recycling of base nutrients, and surface accumulation of forest litter results.

Precipitation infiltrating litter leaches organic compounds (deficient in bases) that in combination become acid.[6] Subsequently, the acid solution percolates through the pedon, first removing exchangeable bases by hydrogen ion (H^+) displacement. The rate of base desaturation varies, depending largely on parent materials, but is slow where a concentration of calcareous or saline ions must be removed. Loss of free lime and/or salts allows clay particles to disperse. Dispersion (also called *peptization*) speeds vertical movement and accumulation in lower horizons. As a consequence, in time the pedon develops stratum of different textures and chemical alteration. The surface horizon is depleted of fine-sized particles, leaving a coarse texture; the subsurface horizon is enriched with fine-sized clay and silt particles. Calcium and magnesium are leached from the topsoil, leaving behind acid silicate minerals, while iron and aluminum may be liberated from their mineral bonding and migrate downward from the surface horizon. These *sesquioxides* (oxides with one and one-half oxygen atoms to every metallic atom) most frequently translocate, or move, in two forms— as *inorganic cations* and as *metal-organic complexes*.

Stobbe and Wright (1959, 162) have summarized these concepts of sesquioxide translocation. Referring to early research on the development of Spodosols, when it was thought that "strong acidity created by the decomposition products of organic matter brought about the solution of sesquioxides (as inorganic cations), which then moved to lower levels where a higher reaction caused their precipitation," they say that the pH level of these soils is normally too high for the solubility of ferric iron. Given reducing conditions, however, where the solum is saturated with water for an appreciable portion of the year, ferrous iron may remain in solution and migrate down the profile as ferrous ions, subsequently oxidizing during periodic dryness. That situation may account for sesquioxide translocation as inorganic cations, at least in poorly drained and high water table sites.

The metal-organic complexes are thought to develop from organic acids interacting with sesquioxides to form soluble products. These migrate with percolating soil solutions to lower horizons. Subsequent processes may lead to precipitation of the sesquioxides within the illuvial layer and

---

[6] Fulvic acids are organic substances that will remain in solution after acidification of a dilute alkali extract taken from the soil.

form a spodic horizon. Several explanations have been offered to describe these processes. Any of the following is possible:

1. Microorganisms destroy the complex organic mineral compounds that initially mobilized the Fe and Al.

2. Insufficiency of infiltrating rainfall fails to carry colloids and solutes further down the profile and forces precipitate occurrence in the illuvial horizon, where dehydration occurs during dry periods.

3. Sieving action by soil particles can lead to clogging of pore spaces by colloids.

4. Negative charges on clay films immobilize positively charged Al and Fe ions.

5. Hydrolysis of the organic-metal complex is induced by changes in the pH level.

The spodic horizon is most readily recognized by its color and structure. The upper boundary is abrupt, and hues, values, and chroma change markedly with depth within a few centimeters. The lowest color values, reddest hues or highest chromas, are always in the upper part of the horizon. Texture is most commonly sandy, coarse-loamy, or coarse-silty, with structure that is either absent or present as crumb, granular, platy, or weak blocky prismatic. The spodic horizon must be at least 2.5 centimeters thick and must have a relatively high cation-exchange capacity. Commonly, it will also exhibit a secondary maximum of organic matter concentration. Since the spodic horizon is an illuvial layer and accumulates minerals removed from the soil's epipedon, an impoverished mineral zone necessarily forms immediately above it. This eluvial layer (E horizon) frequently has the "classical" feature of the originally identified podzol soil, or what is now called the *albic horizon*. If an albic horizon is present, it has light color and relatively coarse texture. The coloration mainly comes from the silaceous sand and silt particles rather than from coatings of other minerals that migrate downward.

Table 9.1 is a profile description of a Spodosol of the suborder Aquod. It is known as a Typic Sideraquod and shows an appreciable accumulation of iron in the spodic horizon relative to organic carbon. This soil, in Curry County, Oregon, formed in beach deposits under a vegetation of pine, bracken fern, grasses, and sedges. The mean annual precipitation of about 1,670 millimeters comes mostly in winter. The soil is saturated in winter months. The mean annual temperature is about 10.4°C (50.8°F), and the mean July temperature is about 15°C (59°F). Slopes are very gentle.

Table 9.1   Profile Description of a Typic Sideraquod

Horizon	Depth (Cm)	Description
O	6-0	Very strongly acid muck with some needles, leaves, and grass litter.
A	0-10	Very dark gray (10YR 3/1) mucky fine sandy loam, dark gray (10YR 4/1) when dry; moderate, fine subangular blocky structure parting to moderate fine granular structure; soft, very friable, slightly plastic; many moderately coarse pores; many roots; very strongly acid, pH 5.0; gradual wavy boundary.
E	10-46	Gray (N5/0 or 6/0) fine sandy loam, white (N6/0 or 8/0) dry; massive; hard, firm, slightly plastic; few fine pores; many roots; strongly acid, pH 5.2; abrupt wavy boundary. [a]
B21h	46-56	Dark reddish brown (5YR 2/2) mucky loam, dark reddish gray (5YR 4/2) dry; weak medium subangular blocky structure or massive—some cementation; soft, very friable; many fine and moderately coarse pores; common roots; very strongly acid, pH 5.0; abrupt wavy boundary.
B22ir	56-81	Yellowish brown (10YR 5/4 and 5/6) and dark reddish brown (5YR 3/4) loamy sand; common medium and coarse dark reddish brown (5YR 3/4) mottles; massive, very hard, very firm; few fine pores; medium acid, pH 5.4; clear wavy boundary.
B3	81-122	Yellowish brown (10YR 5/4) loamy fine sand; few medium distinct strong brown (7.5YR 5/6) and few fine prominent dark reddish brown (5YR 3/4) mottles; massive; slightly hard, firm; many coarse pores; medium acid, pH 5.6; gradual wavy boundary.
C	≥122	Pale yellow (2.5YR 8/4) loamy fine sand; yellowish brown mottles; massive; slightly hard, loose; many coarse pores; reddish brown concretions; medium acid, pH 6.0.

[a] See Appendix III for an explanation of soil color designations.

Spodosols are divided into four suborders (Figure 9.4):

1. *Aquods.* These are wet soils with free iron in very small amounts. The brown and reddish brown colors of the humus in the spodic horizon tend to mask other colors. They range in location from arctic margins to the equator.

2. *Ferrods.* Not a soil found in the United States, its features include a considerable accumulation of iron and a relatively high base-exchange capacity.

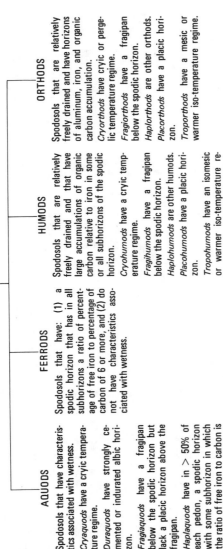

SPODOSOLS

Mineral soils that have a spodic horizon with an upper boundary within 2 meters of the surface or that have a placic horizon cemented by iron that rests on a fragipan or on an albic horizon that rests on a fragipan, and that meets all requirements of a spodic horizon except thickness.

AQUODS

Spodosols that have characteristics associated with wetness.

Cryaquods have a cryic temperature regime.

Duraquods have strongly cemented or indurated albic horizon.

Fragiaquods have a fragipan below the spodic horizon but lack a placic horizon above the fragipan.

Haplaquods have in > 50% of each pedon, a spodic horizon with some subhorizon in which the ratio of free iron to carbon is < 0.2.

Placaquods have a placic horizon that rests on a spodic horizon or a fragipan, or on an albic horizon that is underlain by a spodic horizon.

Sideraquods are other aquods.

Tropaquods have a mean annual soil temperature of 8°C or higher and a mean annual range in soil temperature of less than 5°C.

FERRODS

Spodosols that have: (1) a spodic horizon that has in all subhorizons a ratio of percentage of free iron to percentage of carbon of 6 or more, and (2) do not have characteristics associated with wetness.

HUMODS

Spodosols that are relatively freely drained and that have large accumulations of organic carbon relative to iron in some or all subhorizons of the spodic horizon.

Cryohumods have a cryic temperature regime.

Fragihumods have a fragipan below the spodic horizon.

Haplohumods are other humods.

Placohumods have a placic horizon.

Tropohumods have an isomesic or warmer iso-temperature regime.

ORTHODS

Spodosols that are relatively freely drained and have horizons of aluminum, iron, and organic carbon accumulation.

Cryorthods have cryic or pergelic temperature regime.

Fragiorthods have a fragipan below the spodic horizon.

Haplorthods are other orthods.

Placorthods have a placic horizon.

Troporthods have a mesic or warmer iso-temperature regime.

Figure 9.4 Suborders and Great Groups of the Soil Order Spodosol.

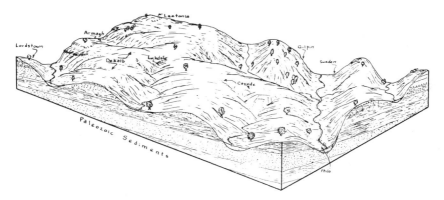

Figure 9.5 Northern Appalachian composite soil associations—Clarion and Potter Counties, Pennsylvania (Unpublished manuscript of classroom exercises, courtesy of Neil E. Salisbury).

3. *Humods.* These are high in accumulated humus, with little or no iron.

4. *Orthods.* The most common Spodosols in northern parts of Europe and North America. They have formed mainly in coarse, acid Pleistocene or Holocene topsoils and usually exhibit organic, albic, and spodic horizons, with or without a fragipan.

Figure 9.5 provides both a landscape description of the northern Appalachian's Allegheny Plateau and also the locations of its dominant soil series. Two series, the *Leetonia* and *Sweden,* are Spodosols. They represent approximately 20 percent of the land area and are both classified as Haplorthods.[7]

The *Leetonia* series formed on broad divides and within coarse sandstones and conglomerates which provide for an acid environment readily amenable to accentuate podzolization processes. Pedon traits are classic. A thin, humus-rich A horizon overlies a thicker, structureless, bleached E horizon (albic horizon). Its light color reflects removal by leaching and eluviation of practically all iron, aluminum, and soluble bases. Organic colloids, clays, iron, and aluminum have been translocated to the B horizon, while bases have washed from the pedon by circulating groundwaters. A spodic horizon is formed in the upper portion of the B horizon as a result of precipitated iron compounds that migrated to this soil zone. The spodic horizon forms a cemented pan that is exceedingly

---

[7] *Haplorthods* are relatively freely drained and have horizons of aluminum, iron, and organic carbon accumulation.

resistant to plant root penetration or movement of fluids. Its brown color indicates that organic colloids have also illuviated to the B horizon.

Sweden soils are minor in areal extent and are found developed within colluvium of valley side slopes. They have a well-developed B-horizon containing iron hydroxides, yet they lack a clearly defined albic horizon.

Other associated soils include Inceptisols (Philo, Lordstown, and Lickdale series) and Ultisols (Armagh, Cavode, and Gilpin series).

LAND UTILIZATION AND MANAGEMENT PROBLEMS

Spodosols provide a life-supporting medium for the world's largest contiguous expanse of boreal vegetation forms. However, trees of construction timber quality also thrive in the more southern regions of the zone. One such area was the famous "Lake Forest" of North America, which once extended from Minnesota to the New England coasts. Only remnants remain of this magnificent forest, which at the time of colonization consisted primarily of white and red pine and eastern hemlock with heights exceeding 60 meters. The excellent character of the forest led to its early and rapid exploitation, a process which was fairly complete by the end of the nineteenth century.

After the loggers had removed the timber, extensive acreages of Spodosols were settled by farmers, who were most often unaware that these highly infertile soils could not support the type of agriculture that takes nutrients out of the soil without replacing them. The result was considerable despair to farmers, whose production costs in conditioning these croplands were so high that general farming became prohibitive, and the land was frequently abandoned.

Within the boreal forest proper, extensive areas have been cleared by the demand for pulpwood and also by fire. Many almost pure stands have been exploited on a large scale because the trees are desirable for making into newsprint. Besides the careless (in many cases) destruction of these magnificent forests, another concern is that these large clearings do not quickly revert to their former state after cutting. Instead, species of birch and aspen are the first to return. These deciduous trees normally produce large quantities of very light seeds capable of being carried tremendous distances by wind. In addition, the seeds can germinate in areas where they are exposed to the dessicating (drying) effects of the sun and wind—much more successfully than do conifer seeds. As a result, instead of deciduous trees occurring as isolated plants or in small clumps, as they did in the original forest, they now occupy extensive areas where clear-cutting or burning has taken place.

A serious concern of land managers is erosion that occurs on logging roads. Ski trail erosion in areas devoted to recreational activities also creates land-use problems. Other than pulpwood and sawlogs, Spodosols of the United States also support syrup-producing maple trees and are cultivated for production of Christmas trees.

Forestry has been the dominant economic activity on the Spodosols, but agriculture is of substantial importance as well. Dairy farms, crop cultivation, hay fields, and pasturelands are all significant parts of the Spodosol landscape. Successful cultivation in a demanding environment and on soils of inherently low fertility is the result of wise land-use practices and modern technology.

With surplus rainfall, the continued washing of the profile depletes the pedon of calcium, magnesium, and nitrogen—essential nutrients for cereal grains and pasture grasses—and induces low base saturation, in some horizons less than 10 percent. This leaves a soil whose exchange complex is dominated by the hydrogen ion (H^+) and which is acid throughout. Due largely to their coarse texture, Spodosols normally have a low capacity to store water. To utilize these soils for agriculture, their limitations must be recognized and *amendments* provided.[8] The earliest attempts to make these soils productive involved incorporating crushed marl and organic matter (in the form of animal manure and root residues) into the soil. The effect of this process, called *marling,* was twofold: (1) The liberation of bases, especially calcium, from the marl reduced soil acidity. Increased pH had the effect of limiting the peptization of clays and encouraging their flocculation (forming into loose clusters). Hence, further translocation of clay was retarded and soil structure was improved. (2) Nitrogen content and soil water-holding capability increased as organic matter was incorporated within the plow layer of the solum.

Marling is not practiced today; instead specially treated limestone is first crushed, sized, and calcinated. Calcination involves heating the limestone at a temperature of about 871°C (1600°F) to burn off carbon dioxide. The resultant product, CaO, is known as quicklime. It will readily react with soil water to produce a calcium hydroxide ($CaO + H_2O = Ca(OH)_2$), which neutralizes soil acidity by replacing exchangeable hydrogen (H) and converting it by recombination into water (H_2O). The importance of limestone as a soil amendment in acidic soils is

---

[8] An *amendment* is any material, such as lime, gypsum, sawdust, or synthetic conditioners, that is worked into the soil to make it more productive. Strictly, a fertilizer is also an amendment, but the term is most commonly used to refer to added materials other than fertilizers.

illustrated in Table 9.2, which shows the yield increase of crops provided with two tons of limestone per acre. These increases amount, in some cases, to as much as 33 percent.

In addition to lime, individual plants have specific nutrient requirements, and the land manager who supplies them in appropriate amounts is the one to achieve maximum crop yields. Dairy farmers cultivating Spodosols in the United States, for example, usually have a rotation that includes small grains, hay, and pasture. The hay consists primarily of red clover, timothy, and alfalfa, all of which require lime and high potash fertilizers to maintain productive stands. These soils also tend to be low in phosphorus and nitrogen. Top dressings of nitrogen, phosphate, and potash mixtures have produced remarkable responses in growth and yields of hays and grains, and have increased pasture production by as much as 300 percent.

The Spodosols and their associated climate regions are well suited for root crops, especially potatoes and sugar beets. Potatoes are very susceptible to their environment. They grow well in acid soils with a pH of about 4.4 to 5.4. If the pH falls much below 4.4, soluble manganese increases in the soil and may become toxic for potato plants. In that case it may be necessary to apply finely ground dolomitic limestone in sufficient quantity to raise the pH level and reduce the manganese problem. If the pH level is raised too high, the plants will experience scabbing. Potatoes also require careful control of available soil phosphorus, potassium, nitrogen, and copper. Boron, magnesium, and sulphur are deficiencies that also may have to be overcome if plants other than potatoes are grown in this area.

Table 9.2   Yield Increases of Selected Crops After an Application of Two Metric Tons of Limestone per Six Year Rotation [a]

Crop	Yield/increase per hectare
Corn (silage)	2.2 metric tons
Oats	5.0 hectoliters
Wheat	6.8 hectoliters
Hay	4.0 metric tons

[a] Data are for New York State.

Source: N. C. Brady, R. A. Struchtemeyer, and R.B. Musgrave, *The Northeast,* The 1957 Yearbook of Agriculture: Soil (Washington, D.C.: U.S. Government Printing Office, 1957), p. 604.

Some of the world's best-known nurseries and market gardens, such as those in the western Netherlands, southeast England, Denmark, and northern Germany, occur on Spodosols. Many years of deep plowing, use of fertilizers, and heavy manuring have created soils that are highly productive for human needs. These sandy soils lack the sticky character of clay soils, and their ease of cultivation has earned them the reputation of being "light." Since the soils are also highly permeable and contain relatively large pore spaces, plants can produce abundant fibrous roots with relative ease. This porous character can be both an advantage and a disadvantage. Sandy soils have a low capacity to store water; hence they warm up quickly in the spring and give plants an early start. On the other hand, drought periods have a rapid impact on vegetation that is sustained on sands. This frequently means the additional expense of supplemental irrigation for high value crops. The intensively utilized Spodosols require a considerable capital investment to ensure maximum production, and they are profitable only where a market is available and the demand for produce is sufficient.

Alfisols and Ultisols

10

Alfisols and *Ultisols* are mineral soils of humid climatic realms. They share the common characteristic of an argillic (clay) horizon, yet differ appreciably in their base status. Alfisols usually have moderate to high base saturation and an ochric epipedon. The Soil Survey staff recognizes six diagnostic epipedons—mollic, umbric, anthropic, plaggen, histic, and ochric. The first five have soil properties that meet specific quantifiable limits. Ochric epipedons, on the other hand, are surface horizons that lack sufficient development to be classified as one of the first five—they are either too light in color; too high in chroma; too low in organic matter; have too high an *n* value; or are too thin to be mollic, umbric, anthropic, plaggen, or histic.[1]

Ultisols (Latin *ultimus,* last) are more thoroughly weathered and have experienced greater mineral alteration than any other midlatitude soil. In addition, extensive leaching has resulted in low base status either in or immediately below the argillic horizon. Yet, the overall gross morphology and horizon sequence of the Ultisols and Alfisols are similar. This has led some researchers to believe that with time and further weathering, Alfisols may eventually degenerate into Ultisols.

Alfisols are found in both middle and tropical latitudes. In the middle latitudes they commonly occur between the Mollisols of the semiarid and subhumid climates and the Spodosols and Ultisols of the humid regimes.

Tropical locations are normally in transitional areas between the Aridisols of the desert and the Ultisols and Oxisols of humid climates (Figures 10.1 and 10.2). Ultisols have a very simple distribution pattern,

---

[1] The *n* value refers to the ratio between the water percentage under field conditions and the percentage of inorganic clay and humus. The *n* value is helpful in predicting whether the soil may be grazed by livestock or support other loads, and the degree of subsidence that would occur following drainage.

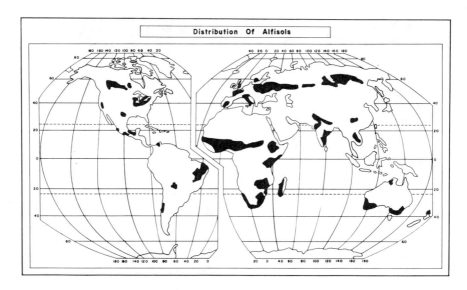

Figure 10.1 World distribution of Alfisols.

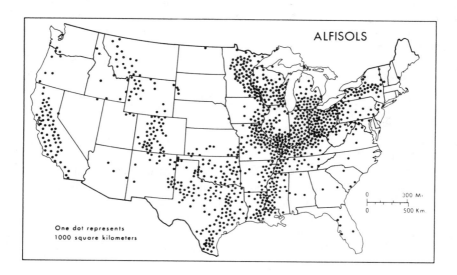

Figure 10.2 Alfisols of the United States. (From Philip J. Gersmehl, "Soil Taxonomy and Mapping," *Annals of the Association of American Geographers,* 67, September 1977, p. 426. By permission of the Association of American Geographers.)

being primarily confined to humid subtropical climates and some relatively youthful tropical landscapes (Figure 10.3 and 10.4).

No other soil order with mature profile development, occurring on an extensive scale, exists in as many diverse climatic and vegetation environments as do the Alfisols. These soils are recognized in micro to macro thermal regimes, in moisture realms ranging from humid to seasonally arid, and under flora that varies from broadleaf deciduous trees to thorn trees and tall savanna grass. Figures 10.5 and 10.6 show two diverse environments in which Alfisol development is dominant. Each example depicts the moisture status of a recognized humid climate, yet each shows only a seasonal period during which some degree of soil desication is expected.

The temperature ranges of the two locations vary considerably, from 6.4°C (11.5°F) at Boromo to 39.2°C (70.6°F) at Fort Nelson. The maximum and minimum mean monthly temperatures also differ noticeably. Boromo has a temperature low of 25.5°C (77.9°F) and a high of 31.9°C (89.4°F), while the monthly minimum at Fort Nelson is −22.4°C (−8.4°F) and the maximum is 16.8°C (62.2°F).

Most Ultisols of the middle latitudes lie south of the Alfisols in a humid subtropical climate. In contrast to the environment of most Alfisols, winters are milder, the temperature regime is usually less continental, and precipitation is greater (Figure 10.7).

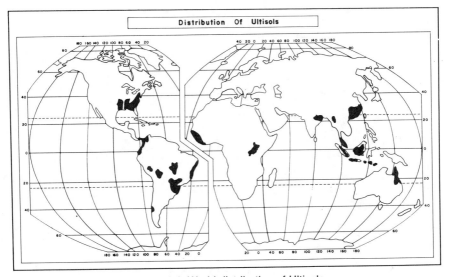

Figure 10.3 World distribution of Ultisols.

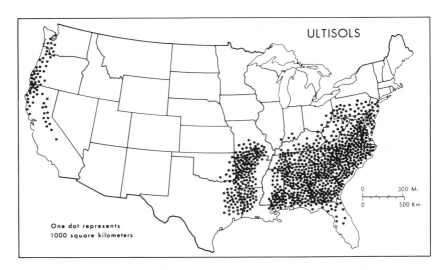

Figure 10.4 Ultisols of the United States. (From Philip J. Gersmehl, "Soil Taxonomy and Mapping," *Annals of the Association of American Geographers,* 67, September 1977, p. 426. By permission of the Association of American Geographers.)

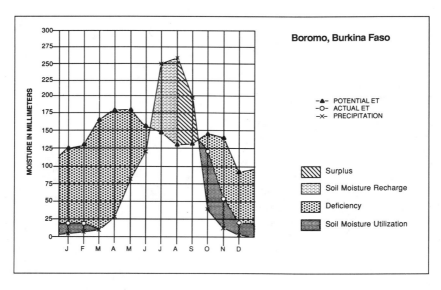

Figure 10.5 Water budget of Boromo, Burkina Faso (based on normal data).

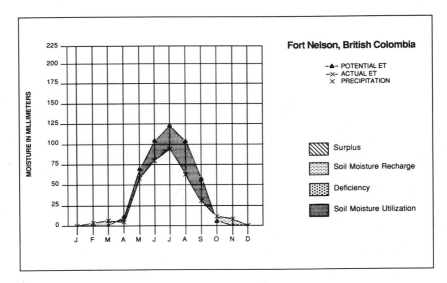

Figure 10.6 Water budget of Fort Nelson, British Colombia (based on normal data).

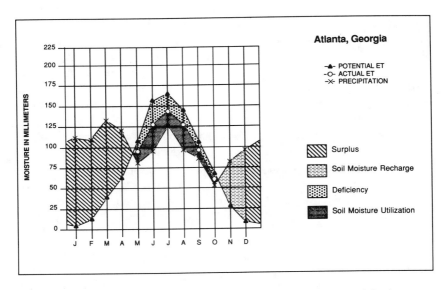

Figure 10.7 Water budget of Atlanta, Georgia (based on normal data).

PEDOGENESIS

Unlike the soils of the moisture-deficient regions, where calcification is dominant, or the subarctic, where podzolization is active, most Alfisols and Ultisols occur in fairly dependable thermal and moisture regions that are not subject to major climatic extremes.[2] The pedon's characteristics result from interaction of the "distance-modified processes" attributed to all earth's major pedogenic regimes.

In midlatitudes, these soils seem to be associated with relatively stable landscapes in which soil water has eluviated and illuviated clays more rapidly than erosion has truncated the surface horizon. The argillic horizon (as we know, a subsurface horizon with an accumulation of translocated clay) and its associated base status, although present largely because of the action of percolating soil water, are also products of long-term influences of vegetation and climate.

The evidence of the presence of an argillic horizon is closely related to processes that were instrumental in its formation, and specifically rests on the character and orientation of the transported clays. Since there is a strong similarity in the fine clays of both the eluvial and illuvial horizons, it is assumed that the clays migrate downward rather than form as decomposition products being synthesized into clays *in situ*. Since water is the moving agent, translocated clay tends to form coatings of oriented clay particles on the channels through or into which water moves or in which it stands. These channels are principally crevices bound by the cleavage faces of peds and pores left by roots or animals. When they are first moved and deposited, the clays tend to be oriented with their long axes parallel to the surface upon which they adhere, and they have the appearance of a surface coating smeared upon ped faces.[3] The formation of this diagnostic horizon requires certain preexisting conditions:

1. Either the parent material must contain very fine clays or subsequent weathering must be capable of producing them;

2. These very fine clays must be subject to dispersion. The presence of certain mineral salts, such as carbonates and free oxides, will tend to flocculate (cluster) clay particles, making it more difficult to displace them. In such cases, chemical weathering and removal of these salts must precede clay migration.

---

[2] In addition to such situations as a high temperature range or deficient precipitation, uniformity of monthly temperature and/or precipitation also can be considered a climatic extreme.

[3] The thin clay coatings have a variety of names, such as clay skins, clay films, clay flows, and illuvial cutans.

A dry soil with these traits, when moistened, experiences a disruption of its fabric and subsequent dispersion of clay. At this point the eluvial clay can be physically carried by the soil solution to a depth where percolation ceases. This process is described below.

> Water percolating in noncapillary voids commonly is stopped by capillary withdrawal into the soil fabric. During this withdrawal the clay is believed to be deposited on the walls of the noncapillary voids. This would explain why illuvial clay commonly is plastered on the faces of peds and on the walls of pores. Such a mechanism for movement and deposition of clay is favored in several ways by a seasonal moisture deficit. First, as already mentioned, wetting a dry soil favors dispersion of the clay; second, when a soil dries, cracks form in which gravitational water or water held at low tension can percolate; third, the halting of percolating water by capillary withdrawal is favored by the strong tendency for a dry soil to take up moisture [National Cooperative Soil Survey, 1975, 19].

ALFISOLS

Alfisols and Ultisols both have an argillic horizon, yet they differ appreciably in base status. Ultisols tend to be infertile and contain few weatherable minerals capable of releasing base nutrients, whereas Alfisols contain a moderate to high reserve of bases. This difference may result from (1) more youthful unweathered minerals within Alfisol pedons, or (2) bases being added to the Alfisols by wind and/or water action. The threshold base saturation minimum of these soils is 35 percent at a depth of 125 centimeters from the surface.[4]

Regardless of geographic location, climatic regime, or vegetative cover, practically all Alfisols also have an ochric epipedon. This surface horizon may exhibit a wide range of characteristics and is found where conditions are not conducive to the formation of other more-striking and taxonomically more-significant diagnostic epipedons.

A description of an Alfisol profile is shown in Table 10.1. This particular soil is an Albaqualf, which seasonally has ground water perched above a slowly permeable argillic horizon. The soil normally exhibits an abrupt textural change between either the ochric epipedon and albic or

---

[4] Exceptions to the depth value are permitted whenever a lithic or paralithic contact is within 125 centimeters of the surface, or when the mean annual soil temperature is less than 14°C.

Table 10.1   Profile Description of an Albaqualf

Horizon	Depth (Cm)	Description
Ap	0-18	Dark gray (10YR 4/1) silt loam, light brownish gray (10YR 6/2) dry; few fine faint olive brown (2.5Y 4/4) mottles; weak, very fine platy structure parting to fine granular structure; friable; very strongly acid; abrupt smooth boundary.
E	18-38	Light brownish gray (10YR 6/2) silt loam, white (10YR 8/2) dry; moderate, fine platy structure; friable; few fine dark concretions; very strongly acid; clear, smooth boundary (about 17 percent clay).
EB	38-43	Pale brown (10YR 6/3) light, silty clay loam, white (10YR 8/2) dry; few fine distinct yellowish brown mottles; weak, fine, subangular blocky structure; friable; thick, continuous silt coats; strongly acid; abrupt smooth boundary.
B21tg	43-63	Dark grayish brown (10YR 4/2) interiors, dark grayish brown (10YR 4/2) and very dark grayish brown (10YR 3/2) ped faces in the upperpart; silty clay, few fine distinct yellowish brown, brown, and yellowish red mottles; strong fine and medium subangular blocky structure; very firm; thick discontinuous very dark gray clay skins; common fine dark concretions; very strongly acid; clear smooth boundary (about 50 percent clay).
B22tg	63-102	Dark grayish brown (2.5YR 5/2) silty clay with some brown in lower part; few fine dictinct yellowish brown, strong brown, and yellowish red mottles; weak medium prismatic structure parting to moderate medium subangular blocky structure; firm; thin discontinuous clay skins; few fine dark concretions; medium acid; gradual smooth boundary.
B3tg	102-142	Grayish brown (2.5YR 5/2) silty clay loam; common fine yellowish brown mottles; very weak medium prismatic structure; firm; common very dark gray stains on peds; thin discontinuous clay skins; few fine hard concretions; medium to slightly acid.

argillic horizon. This particular pedon is formed from Wisconsin loess in Appanoose County, Iowa, under a former hardwood forest cover.

Note in the profile (1) the increase in clay content from the E to the B2tg horizon, (2) the light-colored A and E horizons, and (3) the presence of clay skins in the B horizons.

The Alfisols are divided into five suborders: Aqualfs, Boralfs, Udalfs, Ustalfs, and Xeralfs (Figure 10.8). The *Aqualfs* (Latin *aqua,* water) are associated with a seasonally fluctuating water table and show signs of wetness, such as mottles and iron manganese concretions larger than 2 millimeters. *Boralfs* (Greek *boreas,* northern) are found primarily in cool or cold climates and have albic horizons. *Udalfs* (Latin *udus,* humid) occur in humid climates with a short moisture-deficient period. They tend to be brownish or reddish in color and are relatively freely drained. *Ustalfs* (Latin *ustus,* burnt) are mostly reddish-colored Alfisols of warm subhumid to semiarid regions. They have an ustic moisture regime (a warm, rainy season), but in most of them moisture moves through the soil to deeper layers only in occasional years. Some have a calcic horizon below or in the argillic horizon if carbonates are in the parent material or in the dust that settles on the soil. The *xeralfs* (Greek *xeros,* dry) are mainly associated with a Mediterranean climate. They become dry for extended periods during the summer, but in an occasional year and in some cases every year, moisture moves through the soil in winter to deeper layers.

ULTISOLS

In contrast to Alfisols, the Ultisols are generally low in fertility due to a meager supply of bases. These are the most weathered of all midlatitude soils and are found on surfaces dominantly Pleistocene or older. Glacial ice did not rest on these soils during the Pleistocene period. Consequently, they have been subjected to weathering processes for a much greater length of time than have the soils of humid regimes on their northern boundaries. Extensive chemical weathering of their pedon has led to a removal of bases; yet contrary to expected increased nutrient availability with depth, the reverse is true. The roots of trees go several meters deep in many soils, and the bases they extract at these depths are eventually returned to the soil surface. Before the bases can be moved very deeply into the soil, they are again taken up by plants. Thus, supply of bases is partly a function of the species of plants. Some collect large amounts of bases, and the soil below one of these may be well supplied during the mature life of the plant. But the maintenance of bases in the surface horizon is at the expense of supplies in deeper horizons.

Once the native plant cover is removed, as when planting crops, the stored nutrients are rapidly lost and potential crop yields decrease dramatically. Only through conscientious fertilization programs can permanent agriculture be practiced on such soils.

ALFISOLS

Mineral soils that have either an argillic or natric horizon, and either a frigid temperature regime or base saturation 50' below the argillic horizon of 35% or more.

AQUALFS

Alfisols that have characteristics associated with wetness.

Albaqualfs have: (1) an abrupt textural change between an ochric epipedon or an albic horizon and an argillic horizon, and (2) low permeability in the argillic horizon.

Duraqualfs have a duripan.

Fragiaqualfs have a fragipan.

Glossaqualfs have an albic horizon.

Natraqualfs have a natric horizon.

Ochraqualfs have an ochric epipedon.

Plinthaqualfs have plinthite as a continuous phase or constituting > 50% of the matrix within some subhorizon.

Tropaqualfs have a mean annual soil temperature >8°C and seasonal soil temperature range at 50cm that is less than 5°C.

Umbraqualfs have an umbric epipedon.

BORALFS

Freely drained Alfisols of cool places.

Cryoboralfs have a cryic temperature regime and an argillic horizon.

Eutroboralfs have a base saturation ≥ 60% in all subhorizons of the argillic horizon, and are dry in some horizon at some time in most years.

Fragiboralfs have a fragipan.

Glossoboralfs are either never dry or have base saturation < 60% in some subhorizon of the argillic horizon.

Natriboralfs have a natric horizon.

Paleboralfs have an argillic horizon deeper than 60" below the mineral surface.

UDALFS

These are brownish to reddish, freely drained Alfisols.

Agrudalfs have an agric horizon.

Ferrudalfs have a discontinuous albic horizon and have iron enriched nodules in the argillic horizon.

Fragiudalfs are other udalfs that have a fragipan.

Fraglossudalfs have tongues of albic materials in the argillic horizon and have a fragipan.

Glossudalfs have tongues of albic materials in the argillic horizon and no fragipan.

Hapludalfs are other Udalfs.

Natrudalfs have a natric horizon.

Paleudalfs have an argillic horizon with a clay distribution such that the percentage of clay does not decrease by as much as 20% of the maximum within a depth of 60" from the soil surface.

USTALFS

Have a ustic moisture regime, an epipedon that is both massive and hard when dry, or a calcic horizon within 60" of the surface.

Durustalfs have a duripan within 40" of the surface.

Haplustalfs are relatively thin, reddish to brownish in color that have no petrocalcic horizon within 60" of the surface.

Natrustalfs have a natric horizon.

Paleustalfs have a petrocalcic horizon within 60" of the surface or a dense argillic horizon.

Plinthustalfs have plinthite forming a continuous phase.

Rhodustalfs have argillic horizons redder than 5YR.

XERALFS

Alfisols that have a xeric moisture regime, or an epipedon that is both massive and hard when dry.

Durixeralfs have a duripan within 40" of the surface.

Haploxeralfs are relatively thin reddish to brownish Xeralfs.

Natrixeralfs have a natric horizon.

Palexeralfs have a petrocalcic horizon within 60" of the surface or a solum thicker than 60".

Plinthoxeralfs have plinthite that forms a continuous phase.

Rhodoxeralfs have argillic horizons with colors redder than 5YR.

Figure 10.8 Suborders and Great Groups of the Soil Order Alfisol.

Two features found in some Ultisols are plinthite and fragipans.[5] *Plinthite* (Greek *plinthos,* brick) is related to warm, humid climates and is most extensively developed in tropical soils. It is an iron-rich mixture of clay and quartz that commonly occurs as dark red or brown mottles and will change irreversibly to ironstone hardpans or irregular aggregates upon repeated wetting and drying. Fragipans are weakly cemented layers that impede the downward percolation of water and extension of plant roots. Water either stands above these pans in soils of level areas or moves laterally along the top of the pan if the soils are sloping.

The Ultisols contain five suborders: Aquults, Humults, Udults, Ustults, and Xerults (Figure 10.9). The *Aquults* (Latin *aqua,* water) are wet Ultisols that experience a fluctuating water table and are usually gray or olive in color. *Humults* (Latin *humus,* earth) are more or less freely drained, humus-rich Ultisols. *Udults* (Latin *udus,* humid) are also relatively freely drained, but are humus poor. They generally have light-colored (grayish) epipedons resting on a yellowish brown to reddish argillic horizon. The *Ustults* (Latin *ustus,* burnt) are relatively freely drained Ultisols of warm regions with high rainfall, but a pronounced dry season. They have little organic carbon and most have reddish colors. *Xerults* (Greek *xeros,* dry) are found in Mediterranean climates. They contain moderate or small amounts of organic matter and have orchic epipedons resting on a brownish to reddish argillic horizon.

Figure 10.10 is a landscape mosaic of Alfisols, Ultisols, Entisols, Inceptisols, and Vertisols. That is a representation of five soil orders (one-half the total), within a one-county area. This should suggest to the reader what has been previously emphasized: When a county is classified solely in relation to its most dominant soil, much information regarding the true nature of soils is lost and may lead to erroneous concepts about soil resources.

Soil differences in Montgomery County, Alabama, can be largely attributed to variation in parent material and topographic factors. The county lies in the Coastal Plain physiographic province, East Gulf section. Parent materials are Cretaceous and Tertiary sediments which are tilted seaward. None of the strata is highly indurated, and their structure is simple. The northernmost shown in Figure 10.10, the Eutaw formation, is comprised of sands and clays, except locally, being covered by alluvial floodplains and terraces of the Tallapoosa River. The Selma Chalk, overlying and the next formation to the south (left on the diagram), is a chalky limestone with small quantities of interbedded chalky clay and sands

---

[5] See Chapter 9 for an explanation of fragipan.

ULTISOLS

Mineral soils of the mid to low latitudes that have a horizon in which there are translocated silicate clays but only a small supply of bases.

AQUULTS

These are Ultisols of wet places, when the ground water is very close to the surface part of the year.

Albaquults have a marked increase in the percentage of clay in the upper part of the argillic horizon.

Fragiaquults have a fragipan.

Ochraquults have a ochric epipedon.

Paleaquults are found on old land surfaces and have thick, mottled argillic horizons.

Plinthaquults have plinthite present.

Tropaquults are found in intertropical regions.

Umbraquults are dark colored.

HUMULTS

These are relatively freely drained, humus rich Ultisols of mid or low latitudes.

Haplohumults have some weatherable minerals in the argillic horizon.

Palehumults are reddish Humults of old stable surfaces, with thick argillic horizons and few weatherable minerals.

Plinthohumults have plinthite.

Sombrihumults have a sombric horizon.

Tropohumults are found in intertropical regions and have thin argillic horizons.

UDULTS

These are relatively freely drained, humus-poor Ultisols of humid climates.

Fragiudults have a fragipan.

Hapludults have an ochric epipedon and a relatively thin argillic horizon.

Paleudults are relatively freely drained Udults on very old stable land surfaces.

Plinthudults contain plinthite.

Rhodudults are freely drained Udults with dark colors throughout.

Tropudults are relatively freely drained Udults and found in intertropical regions.

USTULTS

Ultisols of warm regions with high rainfall but with a pronounced dry season.

Haplustults have a relatively thin argillic horizon and 10% or more weatherable minerals.

Paleustults have thick argillic horizons and few weatherable minerals.

Plinthustults have plinthite present.

Rhodustults have dark or dusky red argillic horizons and dark colored epipedons.

XERULTS

These are relatively freely drained Ultisols of Mediterranean climates.

Haploxerults have thin or moderately thick argillic horizons and/or appreciable amount of weatherable minerals.

Palexerults have thick argillic horizons and few weatherable minerals.

Figure 10.9 Suborders and Great Groups of the Soil Order Ultisol.

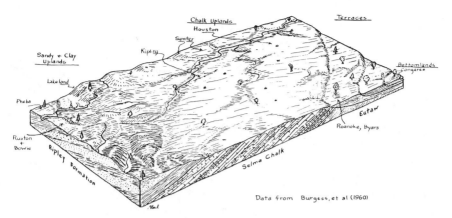

Data from Burgess, et al (1960)

Figure 10.10 Soils of Montgomery County, Alabama (Unpublished manuscript of classroom exercises, courtesy of Neil E. Salisbury).

and occasional clay cappings. It is the least resistant of the area's rocks and has been weathered and eroded into a vale between the parallel cuestas which characterize this section of the Coastal Plain. The southernmost formation is the Ripley, a very sandy collection of rocks with interbedded clays. Both sands and clays are calcareous in some layers and can exhibit signs of induration and produce a relatively erosion-resistant formation—a cuesta—which has been dissected into a belt of hills. The southernmost portion of the Ripley has thick beds of clay.

Ultisols (Bowie, Byers, Pheba, Roanoke, and Rustin series) are relatively deep soils that mainly form in Eutaw and Ripley formations. Differences between them are primarily drainage related. The Udults (Bowie, Pheba, and Rustin) are relatively freely drained and humus poor. Aquults (Byers and Roanoke) are found in depressional, wet areas where the ground water is very close to the surface during part of the year. Alfisols are represented by the Kipling series, which has formed in alluvium from the calcium-rich Selma Chalk. The Houston series of the Selma Chalk upland is a Vertisol. The remaining soil orders are Entisols (the Congaree alluvial soils) and Inceptisols (calcareous Sumter series).

LAND UTILIZATION AND MANAGEMENT PROBLEMS

It is difficult to generalize about land-use patterns and management problems that occur in conjunction with Alfisols and Ultisols. These soils account for approximately 23 percent of the world total,[6] exhibit a wide range in base status, and have considerable variation in length of growing season. Hence, there are many forms of land use on Alfisols and Ultisols when considered in combination at the order level. The basic concept of these soils, however, relates to their development in humid, midlatitude locations. Although they are found in both the tropics and seasonally arid subtropical climates, their utilization and limitations are considered here only in the areas that are characteristic of the basic type.

The Alfisols of the midlatitudes are noted for a relatively high degree of base saturation and fertility. They support some of the earth's most intensive forms of agriculture. In the United States, the well-known agricultural region called the *Corn Belt* occurs mainly on Alfisols and Mollisols. Over the years this area has yielded as much as two-thirds of the nation's corn, oats, and soybeans, and nearly one-half of its alfalfa. This region provides a wide choice of soil and land management systems, including cash grain, dairying, and livestock-feeding operations. Regardless of the system used, lime, legumes, and phosphates give the maximun plant yield.

Because of intensive grain cropping, many soils have lost from one-fourth to more than one-third of their original nitrogen and organic content. With less than 10 percent of the harvested cropland in legume hay and approximately 50 percent in corn, the legumes cannot supply the nitrogen needs of grain crops in rotation. The decrease in available nitrogen in these soils necessitates heavy applications of commercial nitrogen fertilizer. Experiments have shown that 80 to 100 metric kilograms of fertilizer nitrogen per hectare can, on certain Alfisols, double corn yield.

To maintain an economical operation on Alfisols, farm managers not only must be concerned with their soil's nitrogen content, but must also consider (1) the addition of frequently deficient phosphorus and potassium, and (2) economizing on tillage operations. Even with increaseed applications of phosphorus and potassium, the heavy demand for these nutrients and their loss through leaching still pose a problem to farmers who attempt to achieve maximum potential crop yields. Tillage operations present an entirely different set of problems. Excessive tillage results in a decline of soil organic matter, a deterioration of tilth, and an increase in soil erosion. To counteract these problems, some farm managers have

> adopted new methods, which are aimed at minimum tillage for corn in order to save time and money and to maintain better tilth, reduce soil compaction, and enhance water absorption. Among them are substitution

---

[6] Alfisols comprise 14.7 percent and Ultisols, 8.5 percent.

of chemicals for at least one cultivation in the control of weeds; the use of subsurface tillage instead of plowing, with residues left on the surface; and the so-called plow-plant method, in which the corn is planted in the wheel tracks immediately after plowing and the seedbed between the rows is kept loose with minimum tillage [Pierre and Riecken, 1957, 539].

These methods, combined with contour tillage, are especially important on sloping soils that are subject to erosion. A combination of subsurface tillage with a mulch of crop residue can, according to estimates, reduce erosion loss that accompanies traditional plowing and tillage by at least one-half.

Compared with the more northern Alfisols, midlatitude Ultisols have deeper and more thoroughly weathered pedons with much lower base saturation, are acid in reaction, and are relatively infertile. They provide a distinct advantage to the farm manager, however, in that they experience a lengthy growing season (frost-free days range from 200 to 260 in the United States) normally accompanied by abundant rainfall.

Land-use activities on Ultisols are many. The mild climate is favorable to cultivation of crops such as cotton and peanuts that cannot be grown further north. Under proper fertilization programs these soils may also have productive yields of corn, oats, tobacco, and forage crops. Combined with mild winters, forage crops permit the grazing of livestock throughout the year. Beef, dairy cattle, and hogs have become increasingly important land-use functions on Ultisols. Fruits, vegetables, melons, and rice are also of local importance. In addition, the native forest cover represents a huge potential for the pulp and plywood industries, and extensive tracts of land have been set aside for tree plantations.

The same climatic characteristics that make the Ultisol region attractive for agriculture have been largely responsible for the soil's low fertility and mineral deficiencies. Excessive leaching of the profile has led to a removal of minerals from the pedon and a displacement of the bases by hydrogen ions, resulting in a pH that in the United States averages between 5.0 and 5.5. Along with large moisture surpluses is a relatively high temperature regime that prevents the accumulation of significant amounts of organic matter and nitrogen.

Nitrogen is critical to production of healthy crops and must be considered in most rotations. Some land managers plant winter legumes as a cover crop in an attempt to improve soil structure and add nitrogen. However, most farmers prefer to meet nitrogen deficiency through applications of fertilizer. Presented in Figure 10.11 are yield increases for corn, assuming other limiting factors are not present. Other crops respond

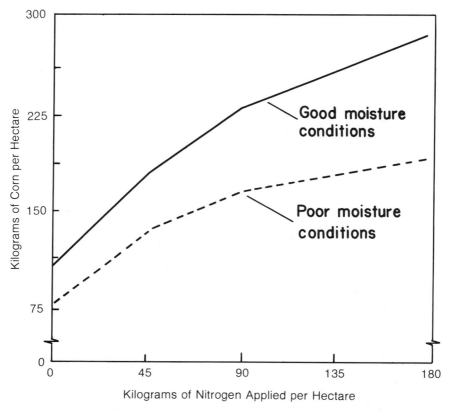

Figure 10.11 Corn yields with nitrogen applications under varied moisture conditions.

in a similar fashion. Pearson and Ensminger (1957), in a report describing the effect of nitrogen applications on forage crops and resultant weight gains of grazing cattle, state that each kilogram of applied nitrogen produced an average 5.9 kilograms of beef.

A well-planned fertilization program not only must provide nitrogen but also must reduce acidity and supply deficient nutrients. Normal practice on most Ultisols is to provide soil calcium, raise the pH level through liming, and supply phosphorus and potassium for most crops. In certain areas where boron, zinc, and sulfur are deficient, their application can also increase crop yields significantly.

Oxisols

11

Oxisols are mineral soils having either an oxic horizon within 2 meters of the surface or plinthite that forms a continuous phase within 30 centimeters of the mineral soil surface. *Oxic horizons* are subsurface horizons at least 30 centimeters thick consisting of mixtures of hydrated oxides of iron and/or aluminum, often in an amorphous state, and containing variable amounts of 1:1 lattice clays. Few weatherable minerals remain in oxic horizons, except for such highly insoluble materials as quartz sand. Hence, there is minimal additional release of bases *via* continued mineral alteration, and the cation-exchange capacity tends to be low.

Plinthite may occur associated with an oxic horizon or by itself. This sesquioxide-rich,[1] humus-poor mixture of clay with quartz changes to ironstone hardpans or irregular cemented aggregates upon repeated wetting and drying. (See Ultisols in Chapter 10 for full explanation of plinthite.)

Oxisols have exhibited the greatest degree of mineral alteration and profile development of any soil. They exist primarily on ancient landscapes in the humid tropics. Seldom found over broad contiguous areas, they are likely to occur in conjunction with more youthful Ultisols, Entisols, and Vertisols. There are two main regions of Oxisol concentration: in South America surrounding the alluvial soils of the Amazon River and in equatorial Africa (Figure 11.1). Less significant areas are found in parts of India, Burma, and Southeast Asia.

---

[1] Remember that a *sesquioxide* is an oxide with one and one-half oxygen atoms to every metallic atom.

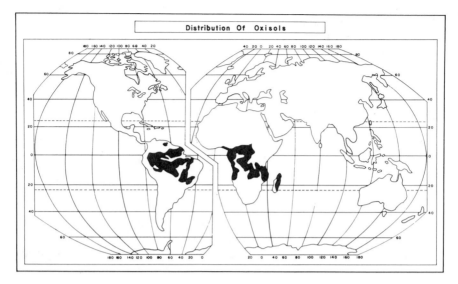

Figure 11.1 World distribution of Oxisols.

CLIMATE AND NATIVE VEGETATION

Climate and vegetation of tropical areas exert a profound interacting influence upon regional pedogenic activity. No other portion of the globe has experienced such a long period of uninterrupted weathering processes and vegetative evolution. Unlike middle and high latitudes where comparatively recent glaciation greatly modified large portions of the surface and restricted development of certain plant species, the more stable tropical climate has encouraged growth of the world's most heterogeneous plant formations, as well as deep and thorough alteration of the mineral crust.

Oxisols are in regions with (1) intense solar radiation, (2) a relatively uniform diurnal (daily) photoperiod (lengths of alternating periods of light and dark), (3) isothermal to near-isothermal monthly temperatures, (4) high potential evapotranspiration rates, and (5) varied amounts of precipitation including seasonal aridity. The extensive range of moisture conditions wherein oxic horizons form has led numerous researchers to conclude that present precipitation patterns are not necessarily responsible for their distribution—many oxic horizons are found in areas of limited rainfall. Rather, it is believed that most extensive areal development of oxic horizons may have taken place under paleoclimates of much higher rainfall.

The humid tropics are noted for large annual receipts of solar radiation. On average the equator receives 2.5 times as much solar radiation as the poles. Yet, seldom do tropical temperature maxima approach those of the midlatitudes. This apparent anomaly is explained by the role of atmospheric circulation and ocean currents. Atmospheric circulation transports away approximately 80 percent of the tropics' surplus energy and prevents its accumulation at the latitude of receipt. Energy transported poleward is in the forms of *latent heat* (utilized in vaporization of water) and *sensible heat* (absorbed as kinetic energy in air masses). Latent heat carried poleward compensates for inequal heating at different latitudes. Energy utilized in the evaporation process serves two additional functions in the hot, humid tropics: (1) Since this portion of solar energy produces a change in the state of water—for example, from liquid to vapor—it does not increase air temperature at the earth's surface; and (2) in converting water into a gaseous form, the atmosphere is enriched with a supply of potential rainfall. Thus, latent heat transport is important not only to moderating temperatures of the tropics but to moisture it receives and makes available for distribution as well.

Due to the energy exchanges taking place under a remarkably uniform photo-period, humid equatorial locations experience little day-to-day variation in temperature. This uniformity induces a monthly moisture demand (PE) that is likewise highly uniform. The water budget graph in Figure 11.2 for Uapes, Brazil, illustrates that this station's monthly PE does not vary from each month's mean of 120 millimeters by more than 13 millimeters in any month.

Total "annual" rainfall in the tropics varies from place to place and fluctuates considerably at the same station as well. When the maximum annual rainfall is expressed as a percentage of the mean, the minimum averages about 60 percent and the maximum around 150 percent. For example, if the mean annual rainfall is 2,540 millimeters, during some years precipitation may be as low as 1,524 millimeters while in others it may be as high as 3,810 millimeters. As far as soil formation is concerned, intensity with which rainfall arrives is as important as the total annual amount. The higher the rainfall is per day, the less value it has to the soil. Within the tropics a daily precipitation event as high as 150 millimeters to 200 millimeters is expected at least once every two years. In Indonesia a daily maximum of 700 millimeters was recorded at Ambon. Obviously, soil cannot hold that much water, and a good portion of the surplus becomes overland flow with high erosive power—particularly on cultivated plots.

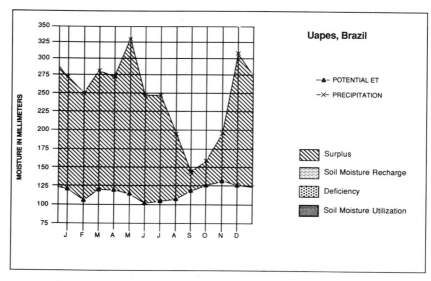

Figure 11.2 Water budget of Uapes, Brazil (based on normal data).

Measurements taken for soils cultivated without precaution against erosion in the Casamance River Basin in West Africa reveal an annual soil loss of approximately 3,750 metric tons per square kilometer on a slope of 1 percent, and 5,880 metric tons per square kilometer on a slope of 1.5 percent, under peanuts, and of 3,290 metric tons and 4,600 metric tons per square kilometer, respectively, under upland rice. Although Casamance lies outside the tropical rain forest climate, the study results indicate the phenomenal potential erosive power within the humid tropics.

In tropical regions where rainfall is heavy and reliable throughout most of the year, a native vegetation called the *tropical rain forest* occurs. Each of the three distinct types—American, African, and Indo-Malaysian—has its own plant families and genera, yet all have evolved plants with similar structures and appearances. Tropical rain forest vegetation is distinctly different from midlatitude forests in both morphology and number of species.

The tropical rain forest is considered the most heterogeneous of all world plant assemblages, with hundreds of different species of trees having been identified within a 1 square kilometer area. The forest also contains the earth's most diverse groups of epiphytes and parasites. The woody plants of this vegetative community are primarily evergreens in that they lose leaves and grow new ones simultaneously. The trees present an array of stratified canopies. Species tend to crown at various heights depending upon sunlight sensitivity (Figure 11.3). Those demanding the most sunlight are normally the tallest and characteristically have broad umbrella-shaped crowns. Lower strata species tend to be less light tolerant and thrive under the shade of their taller neighbors. They are spaced closer and develop conical-shaped crowns. These are mostly thin-barked trees with long straight trunks, few branches beneath the crown, and buttressed bases (for the larger members).

The very complex canopy structure with crowns occurring at distinct levels prevents most solar radiation from directly reaching the surface soil of the forest floor. Only an estimated 1 percent of the energy, reaching the upper-most canopy of a mature stand, is capable of penetrating through the foliage to the soil surface. With such small light intensity, most low-growing plants cannot survive. Therefore, undergrowth of the rain forest, except for mosses and ferns, is relatively absent.

Tropical rain forest vegetation has several important functions related to pedogenesis and soil utilization. It serves as a (1) nutrient reservoir, (2) humus source and food for soil inhabitants (through leaf fall and the like), and (3) protective soil cover. Under conditions of high annual precipitation, the continued leaching of the soil permits rapid loss of soluble base nutrients. The primary way in which bases are conserved in such climates is through temporary immobilization. This is largely accomplished in the rain forest by vegetation, which in the process of

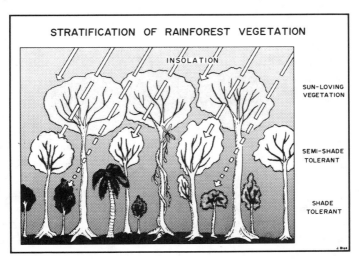

Figure 11.3 Structure of a mature tropical rain forest.

withdrawing life support elements from the soil incorporates nutrients into their cellular structure. Thus, the standing biomass contains a long-term accumulation of nutrients that otherwise might have been lost from the system. Burning this biomass provides a source of fertilizer as organic-mineral complexes are oxidized. Frequently, this is the only form of soil enrichment available for the many isolated fields of subsistence farmers.

Humus cannot exist without an organic source to provide the raw substance from which it originates. Tropical rain forests are well designed to supply a continual and bountiful amount of such material. Rather than a complete seasonal shedding of leaves, evergreens tend to maintain a continuous pattern of shedding older, inefficiently functioning leaves and at the same time replacing them with younger, more vigorous members. Thus, as leaves fall to the ground and aging plants die, the forest floor is supplied with organic litter that, depending upon moisture conditions, is broken down rapidly by microorganisms and insects into various states of decay. Oxidation of organic matter under conditions of good soil drainage is often so rapid that little is left to be incorporated into the soil.

Natural rain forest vegetation protects soil from accelerated erosion and surface dryness. Unprotected surfaces are affected by the high energy impact from raindrops and considerable overland flow, which detach soil particles and wash them downslope at a disproportionately high rate. With a forest cover, raindrop impact is not only reduced, but velocity of surface flow is slowed as well. The canopy also prevents intense heating of soil surfaces, thus reducing soil moisture demand and the possibility for dehydration of epipedon constituents.

PEDOGENESIS

Geomorphic evidence suggests that oxic horizons tend to be associated with surfaces that are mid-Pleistocene in age or older, occur at low elevations (primarily below 1,800 meters), and generally have parent ferromagnesian (basic) rock material. Most of these soils exist on surfaces exhibiting minimal relief—they are normally level or have relatively gentle slopes. Their landscape position is one in which weathered sediments could have been deposited, but recent unweathered sediments could not accumulate on them, and where ground water would not move laterally to them from an area where fresh rock is weathering.

The formation processes leading to development of oxic horizons are called *laterization*. These weathering processes remove silica from the primary silicates, part of the quartz (when present), and alkali and alkaline earth metals. Soluble soil constituents are almost completely washed out by leaching, but some silica can recombine with aluminum to produce alumino-silicate, kaolinite being the final product. Residual weathered silicate is leached out of the pedon. This process is known as *desilication* and consists of silica loss and a simultaneous gain of weathering products of increased stability, among which iron oxides and hydroxides (and at times quartz) are important.

Certain Oxisols that have, or appear to have had, a fluctuating water table contain plinthite (see Chapter 10 under Ultisols). Three distinct stages of plinthite development have been recognized. Stage one is associated with free or nearly free drainage. The atmospheric effects of heating and cooling, and wetting and drying, combined with hydration and hydrolysis, carbonation, oxidation and dissolution, and the interference of soil fauna and flora, all unite to disintegrate and decompose parent rock material. During this first stage, secondary minerals (kaolinite, gibbsite, and geothite) are formed, and bases, silica, and a portion of the aluminum are removed through active leaching processes. As weathering continues, the clay mineral and amorphous sesquioxide colloid content increases, and the erosion base level is approached—leading to the second stage.

In the second stage, rate of water percolation lessens sufficiently to prevent free drainage of all water that infiltrated during the wet season. This results in a seasonal water table within the pedon, even though it may disappear during the dry season. As the soil undergoes repeated appearances and disappearances of the water table, soil-forming processes are characterized by alternating reductive and oxidative conditions. The soil matrix in the fluctuating water table zone will begin to exhibit mottling associated with segregation of iron, Even so, the epipedon may still retain a uniform color.

After continued weathering, the third stage of plinthite formation is initiated and begins as drainage and evapotranspiration become incapable of removing all subsurface water—even in the dry season—and a permanent water table is established. Under water-saturated conditions and low oxygen availability, *gleization* occurs. The primary result is reduction of iron to its mobile ferrous form. A fluctuating water table still exists, even though the highest water level normally is lowered upon weathering. This eventually promotes a thickening of the iron-enriched mottled horizon, identified as plinthite. Portions of the original mottled clay now are solely moistened by capillary rise from the water table, and upper parts actually may become dry. Subsequent drying of the mottled zone can result in cementing of matrix materials through dehydration and crystallization of

Table 11.1   Profile Description of an Oxisol in Mysore State, India

Horizon	Depth (Cm)	Description
A	0-46	Reddish brown (5YR 4/3), silty clay; strong, fine, granular structure; friable; iron concretions; smooth boundary.
B	46-122	Reddish brown (5YR 4/3), clay loam; strong, medium, subangular blocky structure; firm, nonplastic, and nonsticky; bigger iron concretions; diffuse boundary.
C	122+	Reddish brown (5YR 4/3), clay loam; reticulately mottled; strong, medium, subangular blocky structure; firm, slightly sticky; rounded red and black iron concretions; plinthite.

amorphous iron and iron oxyhydrates into cryptocrystalline and crystalline hydroxides and oxides, thus creating an ironstone hardpan.[2]

Table 11.1 describes an Oxisol profile in Mysore State, India.

The B-horizon has all of the characteristics of an oxic horizon, containing a mixture of hydrated oxides of iron and aluminum, 1:1 lattice clays, and insoluble quartz sand. Cation-exchange capacity is low in this soil, and plinthite is present. The plinthite is normally soft when not exposed, but as described, changes irreversibly to ironstone hardpans or irregular aggregates with repeated wetting and drying. In this area dried bricks made from such soil are used for building materials.

Subdivision of Oxisols rests primarily on characteristics that reflect annual and seasonal moisture variation. Moisture is an important variable in tropical soils. It may determine the amount of organic matter, the availability of nutrients, and the degree of plinthite development; it is also its overall index of chemical weathering intensity.

There is a direct relationship between soil moisture status and organic matter content: The latter increases with abundant water and deficient oxygen status. Base saturation, on the other hand, bears an inverse relationship to soil moisture. Under continued leaching, soluble nutrients are removed from the pedon through subsurface drainage. Plinthite, as we know, is related to impeded drainage and a fluctuating ground-water table. All of these activities, except for organic matter accumulation, involve pedochemical reactions with intensities directly related to the soil moisture regime.

---

[2] *Ironstone* refers to plinthite that has irreversibly hardened.

Oxisols are separated into five suborders: Aquox, Humox, Orthox, Torrox, and Ustox (Figure 11.4). The *Aquox* (Latin *aqua*, water) are wet Oxisols having plinthite within 30 centimeters of the surface and are saturated with water at this depth during part of the year. *Humox* (Latin *humus*, earth) are normally found in high altitudes and have relatively high contents of organic matter. Yellowish to reddish in color, the *Orthox* (Greek *orthos*, true) have short or no dry seasons. The *Torrox* (Latin *torridus*, hot and dry) are the Oxisols of arid climatic regimes. *Ustox* (Latin *ustus*, burnt) are red in color and are dry for extended periods, although they are moist for at least ninety days per year.

LAND UTILIZATION AND MANAGEMENT PROBLEMS

In general, tropical soils tend to be of low fertility, with the exception of some that are alluvial or volcanic in origin. Their properties are unique from those of the middle latitudes and require special management techniques to attain their yield potential. Under rain forest cover, fertility of humid tropical soils is maintained in a delicate equilibrium. Organic matter falling from trees is decomposed, releasing nutrients and making them available for plant utilization in a continuous cycle that operates with the same small capital of nutrients. Once rain forests are cleared by man, this balance is destroyed. Soil fertility subsequently declines as oxidation and leaching take place under high temperatures and rainfall without a protective mantle of natural vegetation.

Considering the limited fertility of most tropical soils, they support a surprising number of varied economic activities. Land use ranges from shifting cultivation on subsistence levels to permanent plantations, and from grazing to lumbering operations. By far the most universal use of land throughout most of the humid tropics has been by shifting agriculturists (Figure 11.5).[3]

In an area of shifting cultivation, the landscape consists of extensive forested areas broken by cleared patches where cultivation takes place. There is constant clearing of new fields and abandoning of old ones after two or three years of producing vegetables and grain crops. Figure 11.6 illustrates such a unit in the Republic of Panama. The farm *(finca)* covers approximately 100 hectares, providing sufficient space for regeneration of bush fallow and clearing of new land.

---

[3] *Shifting agriculture* refers to subsistence economic activities and involves the use of fire to clear lands for temporary utilization, such as 2 to 3 years. Names by which it is known include *milpa, swidden, ladang, caingin, roza,* and others.

OXISOLS

Mineral soils that have an oxic horizon within 80″ of the surface or plinthite that forms a continuous phase within 12″ of the mineral surface of the soil.

AQUOX

Oxisols that have plinthite forming a continuous phase within 12″ of the surface; or have an oxic horizon that has characteristics associated with wetness.

Gibbsiaquox have no plinthite forming a continuous phase within 12″ of the surface and have cemented sheets containing ⩾ 30% gibbsite.

Ochraquox have no plinthite forming a continuous phase within the upper 40″ and have an ochric epipedon.

Plinthaquox have an ochric epipedon and plinthite that forms a continuous phase within the soil between 12″ to 50″.

Umbraquox have an umbric or histic epipedon and neither a plinthite formation within 50″ of the surface or gibbsite.

HUMOX

Oxisols that are always moist or have no period when the soil is dry below the surface 7″ for 60 days or more. They have high contents of organic matter and a low supply of bases.

Acrohumox are the most intensely weathered Humox. They have lost virtually all ability to retain bases in their mineral fraction.

Gibbsihumox have within 40″ of the surface, cemented sheets or a subhorizon that contains gibbsite.

Haplohumox have an oxic horizon with a small cation-exchange capacity in their mineral fraction.

Sombrihumox have an oxic horizon with a subhorizon that is darker in color and contains more organic carbon than the overlying horizon.

ORTHOX

Oxisols exclusive of Aquox that have short or no dry seasons. They are most common near the equator.

Acrorthox have lost virtually all ability to retain bases in their mineral fraction.

Eutrorthox are Orthox with relatively high base status.

Gibbsiorthox have gibbsite within 50″ of the surface.

Haplorthox have few bases, but more than the Acrorthox, and the clays have a modest cation-exchange capacity.

Sombriorthox have an oxic horizon with a subhorizon that is darker in color and contains more organic carbon than the overlying horizon.

Umbriorthox are Orthox that have an umbric epipedon or appreciable amounts of organic matter.

TORROX

These soils are usually dry in most years in all parts of the soil. These are the Oxisols of arid climates. (No subdivision of Torrox is presently recognized.)

USTOX

Oxisols that have some subhorizon below the surface 7″ that is dry for at least 60 consecutive days annually. They tend to be found near the Tropics of Cancer and Capricorn.

Acrustox are red or dark red Ustox that have virtually no ability to retain bases in their mineral fraction.

Eutrustox have a mollic or umbric epipedon, an appreciable supply of bases, and moderate to high base saturation in the oxic horizon.

Haplustox are red or dark red Ustox that have clays with modest cation exchange capacity, but have low base saturation in some or all parts of the oxic horizon.

Sombriustox have an oxic horizon with a subhorizon that is darker in color and contains more organic carbon than the overlying horizon.

Figure 11.4 Suborders and Great Groups of the Soil Order Oxisol.

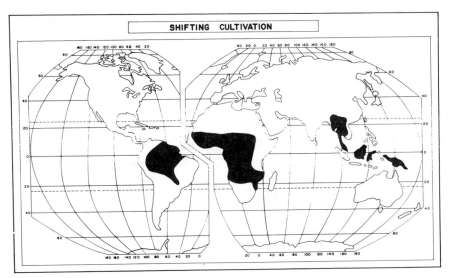

Figure 11.5 World distribution of shifting cultivators.

In preparing land for cultivation, the farmer usually chooses select spots of virgin forest if available. They are easier to clear since the undergrowth is not dense and tangled. In addition, a forest of tall trees indicates an area that has not recently been used for cultivating crops, hence containing a greater reservoir of stored nutrients. Clearing the land normally involves hacking out lianas, saplings, low underbrush, and either girdling or felling trees. This generally takes place during the period of minimum rainfall, thus allowing the vegetative debris to dry and become flammable.

Burning is an essential feature of the shifting agricultural system, for only in this way can the ground be cleared of the tremendous mass of felled vegetation. All plants in the herb layer are destroyed in the burn, leaving the ground clean. Greenland and Nye (1965, 67) give the following comments regarding the effect of burning in these areas:

> All the nutrient elements in the fallow except nitrogen and sulphur are preserved and added to the soil in the ash. The loss of organic carbon and nitrogen that is entailed is often deplored but it is unavoidable. There is no practicable way of incorporating this material in the soil to form stable humus. In grassland, burning is equally essential: a farmer's primary object is to prevent regrowth of the grass by the simplest means, and it is quite impracticable to bury grassland vegetation with a hand hoe.

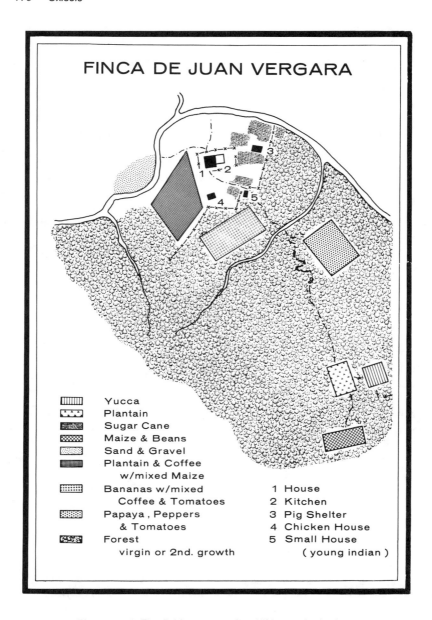

Figure 11.6 The field patterns of a shifting agriculturist.

Large quantities of nutrient ions from the standing vegetation and the litter layer are spead in the ash on the surface of the soil in the form of carbonates, phosphates, and silicates of the cations. Nearly all of the nitrogen is, however, lost to the atmosphere as ammonia, gaseous nitrogen, or other oxides of nitrogen, and sulphur as sulphur dioxide.

... plant growth is better in previously heated than in unheated soils. Any seeds present in the soil during the heating are liable to be killed or to have their germination retarded. The supply of available nutrients in the soil solution is increased by heating, and in particular the rate of nitrogen mineralization is generally greater subsequent to the burn than before it. In most instances there is a profound change in the numbers and composition of the microflora of the soil.

The effect of heat on the microbial population is usually referred to as "partial sterilization," and it is similar to effects produced by drying or by treating the soil with antiseptics. The sterilizing treatment leads to an initial decrease in the microbiological population, after which it redevelops, usually with a modified composition, to a level greater than before. From the soil fertility viewpoint the most important effect of the increase is the change in the rate of nitrogen mineralization. This generally shows parallel changes to the population changes.

After the vegetation has been burned, the soil remains relatively unattended until approach of the rainy season, when crops are planted. The small patches of land upon which crops are grown quite often sustain a variety of intercropped plants, such as maize and beans in the Americas, maize and rice in Africa, and other combinations. In some regions where the rainy season is long enough, two to three crops per year may be harvested.

A second clearing and burning normally take place after the first year of crop growth. This weakens suckers and destroys seedlings. However, it also encourages the growth of grasses.

There is ample evidence to prove that even when starting with land that is either a virgin or a mature secondary forest, crop yields fall fairly rapidly. This does not concern the native farmer. He moves on to another patch of land because encroachment of grasses and the labor of weeding them is uneconomical. Less effort is demanded to clear new land. Encroachment of grasses and weeds is a serious management problem in the tropics. Unlike the grasses of the middle latitudes, which have a high requirement for bases and tend to increase fertility by recycling base nutrients, the tropical grasses tend to be much less effective than forests in preventing nutrient loss through leaching.

As population demands for food increase, pressure for improved crop yields from tropical soils make it necessary for the shifting cultivator to clear plots in shorter periods of time. He does not appear to respond to such pressure by cropping longer. With less time for forest regeneration, the shorter fallow period is conducive to establishment of savanna vegetation. This is accompanied by a decline in the amount of phosphorus mineralized from it. The length of time an area can be cultivated without fallowing averages about three to five years. The fallow period normally is approximately three times the length of the cropping period. This is in general agreement with the time estimated to reach a near-equilibrium organic accumulation on forest floors.

Land-use systems in the tropics have often been condemned as exploitive, squandering the natural resources of the area. Main objections focus upon the loss of forest vegetation, destruction of humus, and loss of nutrients through accelerated leaching. The first objection can be assessed in terms of the demand for agricultural land. Without forest removal, agriculture would be impossible. Second, under an appropriate fallow system humus can be maintained at satisfactory levels. Third, although nutrients are definitely leached out of the topsoil during the cropping period, it has not been determined that they are removed beyond the root zone of the fallow. Therefore, nutrients may still be available for recycling within the biotic system.

Histosols

12

Unlike other soil orders, *Histosols* are not considered primarily mineral, but organic. They are commonly called *bog, moor, peat,* or *muck;* they are last in the list of soil orders, last in areal importance, and have been given the least amount of attention in the more recent soil classification systems. In fact, the National Cooperative Soil Survey team considers its current means of differentiating these soils as "provisional."

Histosols are more than 12 to 18 percent organic carbon by weight (depending on the clay content of the mineral fraction and the kind of materials) and well over half organic matter by volume. Unless drained, most Histosols are saturated or nearly saturated with water a large portion of the year. In certain cases they may be only an organic mat floating on water. Presence of water is the common denominator in all Histosols regardless of location.

These soils can form in virtually any climate (Figure 12.1), even in arid regions, as long as water is available. They occur in the tundra and in the tropics, and their vegetation mantle consists of a wide variety of water-tolerant plants.

In addition to water, controlling factors that regulate accumulation of organic matter include the temperature regime, character of the organic debris, degree of microbial activity, and the length of time in which organic accumulation has taken place. The deposition of organic material within water-saturated environments rapidly leads to depletion of oxygen in the wet zone through decomposing aerobic (oxygen-demanding) microorganisms. This eventually results in an anaerobic (without oxygen) milieu, in which the rate of organic matter mineralization is considerably reduced. As a consequence of the impeded decay process, organic matter continues to accumulate. Development of these soils is, therefore, obviously from the mineral surface upward, characterized by a deepening of the organic layer.

Figure 12.1 Histosols of the United States. (From Philip J. Gersmehl, "Soil Taxonomy and Mapping," *Annals of the Association of American Geographers,* 67, September 1977, p. 427. By permission of the Association of American Geographers.)

The major distinction among Histosol suborders relates to the degree of decomposition of organic detritus and the soil's moisture status. There are four suborders: Fibrists, Folists, Hemists, and Saprists (Figure 12.2).

Fibrists (Latin *fibra,* fiber) are Histosols of relatively unaltered plant remains. The plant debris is so minimally decomposed that it is not destroyed by rubbing, and botanic origins are clearly evident. They may consist of partially decomposed wood or of the remains of such plants as moss, grass, sedge, and papyrus, or of mixtures. The reasons for the preservation of plant remains vary, but absence of oxygen is probably the most important factor. A lowering of the water table, or its seasonal fluctuation, increases the availabilty of oxygen, enhances decomposition, and encourages rapid destruction of fibers. Rate of mineralization varies, however, according to the thermal regime and the nature of the vegetative debris. Bald cypress, for example, has a greater degree of decomposition resistance than most woods, grasses, and sedges. As a result, accompanying drainage, mixed organic deposits will show an increased percentage of volume occupied by the Bald cypress, if this vegetative form is present.

Folists (Latin *folium,* leaf) are relatively freely drained Histosols consisting of an organic horizon derived from leaf litter, twigs, and branches (but not sphagnum), resting on rock or on fragmental materials

HISTOSOLS

These are soils that are dominantly organic.

FIBRISTS

These are Histosols that are comprised of organic matter that is decomposed. The botanic origin of the plant remains can be readily determined.

Borofibrists are found in frigid temperature regimes.

Cryofibrists are found in colder regimes than the Borofibrists.

Luvifibrists have *humilluvic* materials.

Medifibrists are found in the middle latitudes.

Sphagnofibrists are mainly derived from various species of Sphagnum and associated herbaceous plants.

Tropofibrists are found in the intertropical areas.

FOLISTS

These are organic soils consisting of leaf litter, twigs, branches, resting on rock or fragmental mineral materials.

Borofolists are found in frigid temperature regimes.

Cryofolists are found in colder regimes than Borofolists.

Tropofolists are found in intertropical areas.

HEMISTS

These are organic soils in which the botanical origin of up to two-thirds of the materials cannot be determined.

Borohemists are found in frigid temperature regimes.

Cryohemists are found in colder regimes than the Borohemists.

Luvihemists have humilluvic materials.

Medihemists are found in the middle latitudes.

Sulfihemists have sulfidic materials within 40″ of the surface.

Sulfohemists are acid sulfate soils.

Tropohemists are found in the intertropical regions.

SAPRISTS

The soils consist of almost completely decomposed plant remains.

Borosaprists are found in frigid temperature regimes.

Cryosaprists are found in colder regimes than the Borosaprists.

Medisaprists are found in the middle latitudes.

Troposaprists are found in the intertropical regions.

Figure 12.2 Suborders and Great Groups of the Soil Order Histosol.

made of gravel, stones, and boulders, with the interstices filled or partially filled with organic materials. There is very little evidence of mineral soil development, and plant roots are restricted to pockets of organic material.

Following is a description of a *Borohemist* (a Hemist of a frigid thermal regime). This soil, located in Chase County, Michigan, is in an undrained depression of a sand glacial outwash plain.

Table 12.1   *Profile Description of a Borohemist (Chase County, Michigan)*

Horizon[a]	Depth (Cm)	Description
Oe1	0-13	Dark reddish brown (5YR 3/3), dark reddish brown (5YR 3/2) rubbed hemic materials; about 55 percent fiber, about 25 percent rubbed; weak thick platy structure; nonsticky; about 60 percent of the fibers are sphagnum mosses and the remaining portion is herbaceous; extremely acid; abrupt, smooth boundary.
Oe2	13-48	Very dark brown (10YR 2/2) on broken face and rubbed hemic materials; about 40 percent fibers, 15 percent rubbed; weak thick platy structure; nonsticky; 85 percent of fibers are herbaceous, the remaining woody or mosses; extremely acid; clear smooth boundary.
Oe3	48-71	Dark brown (10YR 3/3) on broken face and rubbed hemic materials; about 50 percent fiber, 25 percent rubbed; weak thick platy structure; nonsticky; 85 percent of fibers are herbaceous with remaining portion woody; few woody fragments 2.5 cm to 15 cm in diameter; extremely acid; clear smooth boundary.
Oe4	71-152 +	Dark reddish brown (5YR 3/3) on broken face and rubbed hemic materials; about 60 percent fibers, 30 percent rubbed; massive; nonsticky; about 90 percent of fibers are herbaceous with a few woody fragments; very strongly acid.

[a] O is used to designate an organic horizon; *e* (Oe) represents a Hemic horizon; other organic horizons include *i* (Oi) for fibric, and a (Oa) for sapric.

Hemists (Greek *hemi,* half) are Histosols in which the organic material is strongly, but not completely, decomposed. Enough, however, has been broken down to the point that its biologic origin cannot be determined and/or the fibers can be largely destroyed by rubbing between the fingers. These soils are normally found where the ground water is at or very near the surface most of the year, unless artificially drained. Ground-water levels can fluctuate, but they seldom drop more than a few centimeters below the surface tier. These soils were once classified as bog. They have been identified from the equator to the tundra, usually in closed depressions and in broad, very poorly drained flat areas such as the coastal plains.

Saprists (Greek *sapros,* rotten) consist almost entirely of decomposed plant remains. The botanic origin of the material is, for the most part, obscure, and their color is usually black. They occur in areas where the ground-water level fluctuates within the soil and are subject to aerobic decomposition. When Fibrists and Hemists are drained artificially or naturally, they will normally continue to decompose and convert to Saprists.

LAND UTILIZATION AND MANAGEMENT PROBLEMS

Where Histosols are drained in such a way as to permit the rapid removal of excess water, yet retain the water level at a relatively shallow depth, they can be utilized very profitably for intensive forms of crop production. The wise management of water is critical. In their original state, these soils are too wet either to support the operation of farm equipment or to produce crops. When they are drained, they become aerobic and may oxidize and subside rapidly. Rate of oxidation and subsidence is very closely linked with the thermal regime in which a particular Histosol occurs. In Florida, subsidence takes place at an estmated rate of 5 to 8 centimeters each decade for Histosols under cultivation. These rates are significantly reduced farther poleward where temperatures are lower.

The ease with which plant roots can penetrate, as a result of low bulk density, permits intensive cultivation of a variety of crops on Histosols. The major agricultural limitation in any given area is climate. A late spring or early fall frost in the midlatitudes can create severe economic losses and make it unadvantageous to produce certain crops.

Cabbage, carrots, celery, cranberries, mint, onions, potatoes, and a variety of root crops have been successfully cultivated on these soils where managed properly. In addition to sophisticated drainage systems, potential crop production can be realized only when the farm manager follows a conscientious fertilizer program and protects soil from erosion and fire hazards. Organic soils normally require little or no nitrogen fertilization, but they do need relatively heavy applications of phosphorus and potassium. In addition, calcium and magnesium are frequently in short supply and must be compensated for through liming. Deficiencies of secondary elements depend largely upon the crop cultivated and the character of the organic debris. The more common deficiencies that occur include copper, zinc, and boron (especially for celery and clover).

When Histosols become dry through excessive drainage or during prolonged drought periods (when the water table drops), they are subject to fire damage. The dried organic material is a virtual tinderbox that can be ignited by careless human acts or by naturally occurring phenomena such as lightning. Once a fire is started, these soils may smolder for months. The degree of soil destruction from uncontrolled fires can be extensive. In fact, a current theory regarding formation of Drummond Lake in the Dismal Swamp of North Carolina involves the concept of fire "burning out" a depression in relatively thick deposits of peat. This is not hard to believe, for documents of the U.S. Department of Agriculture record numerous accounts of such destruction in swamps and bogs during periods of intense drought.

Other problems relating to Histosol management include erosion and bearing capacity. The very light character of the partially decomposed organic matter, which makes this soil ideal for plant root development, also contributes to its susceptibility to wind erosion. An intense storm with gusty winds can rapidly remove the surface organic horizon. Many farm managers try to counter this problem by maintaining trees as a windbreak on the margins of their fields. The problem of wind erosion increases significantly whenever the soil is left bare, particularly after harvest and prior to spring planting.

Histosols have very little capacity to support weight. Soil subsidence and compaction under weight cause stresses that can lead to the cracking and rupturing of constructions such as roads and buildings. Therefore, it is frequently necessary to drive concrete pilings deep enough to reach the subsurface mineral strata for building supports, and in the case of roads the entire organic layer may have to be removed and replaced with a substitute such as sand and/or gravel. Similarly, the farm manager must exercise care in using equipment on these soils. The excessive weight of heavy machinery breaks down structure, compacts the soil, and reduces the ability of plant roots to penetrate.

Appendix I: Structure of Clay Minerals

Electron microscopes have been used to magnify (as much as 15,000 times) and photograph clay minerals; even so, it is difficult to examine clays due to the inability to completely separate them as individual particles and because of their extremely small size, which ranges from about 0.002 to 2 microns. As noted in Chapter 1, clay is generally classified into one of two broad groups—the *silicates* and *hydrous oxides*. The most common are known as *silicates,* or more properly *aluminosilicates,* with primary elements being atoms of oxygen, silicon, and aluminum. The building block of these clays is a *silica tetrahedron* and *alumina octahedron.*

The silica tetrahedron is a pyramid-like structure wherein a silicon atom is surrounded by four oxygen atoms, as illustrated below:

A. Silica Tetrahedra   B. A partial Tetrahedral sheet

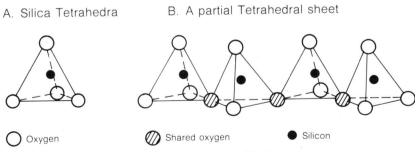

○ Oxygen   ⊘ Shared oxygen   ● Silicon

Figure AI.1 Structure of the silica tetrahedron.

179

Silica tetrahedra share basal oxygen atoms in repeating patterns of six to form hexagonal units that combine to form *silica sheets*. (It should be noted that individual tetrahedra are so minute in size that thousands must be joined together to form the very smallest clay mineral.)

The alumina octahedron is constructed of one aluminum atom surrounded by six oxygen atoms, or hydroxyls:

A. Alumina Octahedron   B. A partial Octahedral sheet

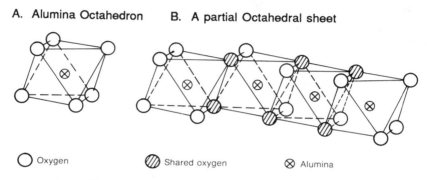

○ Oxygen   ⊘ Shared oxygen   ⊗ Alumina

Figure AI.2 Structure of the alumina octahedron.

Like the tetrahedra, the alumina octahedrons share common bonding with neighbors to produce an alumina octrahedal sheet. The individual tetrahedral and octrahedral sheets interlock in layers to form clay crystals.

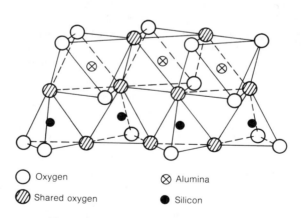

○ Oxygen   ⊗ Alumina

⊘ Shared oxygen   ● Silicon

Figure AI.3 The clay mineral kaolinite.

SILICATE CLAYS

The alumino-silicate clays are grouped into families based upon their layered structures. The *Kandites* possess the simplest structure, having one silicon sheet for each aluminum sheet; *Smectites* are constructed of one aluminum sheet sandwiched between two silicon sheets; *Vermiculites* are similar to the Smectites but have an altered atomic arrangement; and *chlorites* have two silicon sheets, one aluminum sheet, and one magnesium sheet.

Kandites. A family of clays having 1:1 (one-to-one) layered minerals, that is, they are made up of one tetrahedral sheet and one octahedral sheet. They include *kaolinite, halloysite, dickite,* and *nacrite. Kaolinite* has the classic traits of the Kandites, wherein aluminum and silicon atoms are balanced ($Al_2Si_2O_5(OH_4)$), and their respective octahedral and tetrahedral sheets share common oxygen or hydroxyl bonding. These form stacked, platey crystals that readily slide over one another, providing for soil plasticity. The remaining kandites are chemically identical to kaolinite, except *halloysite,* which contains interlayered water molecules within the crystal structure and has a tubular, rather than a platey, shape; *dickite* and *nacrite* each have a slightly different morphology associated with the angular orientation of the tetrahedra and octahedra within their respective sheets.

Smectites. A group of clays having a 2:1 (two-to-one) layered structure. Members of this clay family are montmorillinite, beidellite, hectorite, nontronite, saponite, and sauconite. They originate via either vermiculite weathering or the recombination of products released in the degradation of other minerals. All smectites have the same general structure: an aluminum octahedral sheet bonded between two silica tetrahedral sheets ($Al_4Si_8O_{20}(OH)_4 \cdot H_2O$), wherein the tetrahedron face one another but do not bond. This results in making both space available in the clay interior and also a layered structure that is weakly held together. Consequently, water may enter the crystal, causing the clay to expand or swell. When dehydrated it will shrink and cause soil cracks. A clay's *shrink-swell potential* is a critical engineering consideration and is determined through evaluation of a soil's volume change when wet and dry. Expressed as the *coefficient of linear extensibility* (COLE), it is derived as follows:

$$COLE = \frac{\text{Length of moist soil sample}}{\text{Length of dry soil sample}} - 1$$

When COLE values exceed 0.03, smectite is present; if greater than 0.09, shrink-swell will be sufficient to cause damage to structures. The latter can include broken gas and water lines, broken highway pavement, caved-in basements, and displaced building foundations. Because of their numerous sites to hold cations, however, Smectites have a high cation-exchange capacity (CEC).

Like the Kandites, Smectites may vary in atomic composition while retaining their overall structure. This is called *isomorphous substitution* and occurs when one atom replaces another, such as an aluminum atom substituting for a silica.

Montmorillinite is the most common Smectite and has some aluminum (Al) and/or magnesium (Mg) replacing silicon. Other forms and their silicon substitutes are beidellite (Al), nontronite (Al,Fe), saponite (Al,Mg), hectorite (Mg,Li), and sauconite (Al,Zn,Mg,Fe). *Pyrophyllite* has the balanced chemical formula classic to the smectite family.

Vermiculite. A 2:1 alumino-silicate mineral with extensive isomorphous substitution. In the tetrahedral sheet, considerable aluminum (Al) replaces silicon (Si), whereas iron (Fe^{3+},Fe^{2+}) and magnesium (Mg^{2+}) substitute for aluminum in the octahedral sheet. Within the crystal, water molecules and magnesium frequently occupy spaces not filled by structural atoms. This allows for shrink-swell but not the potential of Smectites. *Chlorites* are considered 2:1:1 layered clays. They are nonexpanding and have a magnesium hydroxide ($Mg_6(OH)_{12}$) sheet (called *brucite)* interlayered between each set of 2:1 layered silicon-aluminum sheets. In some cases aluminum or iron may substitute for magnesium in the brucite sheet. The presence of brucite promotes strong sheet bonds and eliminates significant shrink-swell. *Illite,* also called *hydrous mica,* is a 2:1 structured clay, without definite chemical composition. Usually part of the silicon is replaced by aluminum. In addition, potassium (K) occurs in the interlayers, providing for the stability and nonexpanding traits of illite.

NONSILICATE CLAYS

There are soils containing clay-sized minerals (<2 microns), but in which silicon and/or aluminum are not primary building blocks. These are the sesquioxides (hydrous oxides) of iron or similar metals with varied proportions of water and hydroxyls in their structure. They range in form from crystalline to noncrystalline (amorphous) states. The iron hydrous oxides, in particular, provide soils and water pipes with their rusty colorations of yellow, red, and even dark red-browns. Sesquioxides are oxides chemically having one and one-half oxygen atoms to every metallic atom.

Appendix II: Descriptive Soil Profile Symbols

O	Horizon dominated by organic matter.
A	Organic-rich, mineral horizon at or adjacent to the surface.
E	Mineral horizon of maximum eluviation.
B	Mineral horizon of maximum illuviation and formed beneath an O, A, or E horizon.
C	Weathered parent material.
R	Underlying consolidated bedrock.

The following are recognized transitional horizons:

AB	A horizon transitional between A and B, dominated by properties characteristic of an overlying A horizon.
BA	A horizon transitional between A and B, dominated by properties characteristic of an overlying B horizon.
AC	A horizon transitional between A and C, dominated by properties characteristic of an overlying A horizon. Common in soils lacking a B horizon.
EB	A horizon transitional between E and B, dominated by properties characteristic of an overlying E horizon.
BE	A horizon transitional between E and B, dominated by properties characteristic of an underlying B horizon.
BC	A horizon transitional between B and C, dominated by properties characteristic of an overlying B horizon.

The following additional symbols are used in combination with the previously described horizon designations. These give more detailed information about the composition of a soil horizon:

a   Organic material which is highly decomposed.

b   A buried soil layer.

c   Concretions cemented by materials harder than lime.

e   Organic material at a transitional stage of decomposition.

f   Frozen ground.

g   A water-logged (gleyed) layer.

h   An accumulation of illuvial humus.

i   Slightly decomposed organic matter.

k   An accumulation of calcium carbonate.

m   An indurated layer, or hardpan, due to silication or calcification.

n   Accumulation of sodium as an exchangeable ion.

o   Accumulation of residual sesquioxides.

p   A layer disturbed by plowing.

q   Accumulation of silica.

r   Weathered bedrock.

s   An accumulation of illuvial iron.

t   An accumulation of illuvial clay.

v   Plinthite.

w   Color development where illuvial material is absent.

x   A fragipan.

y   An accumulation of gypsum.

z   An accumulation of soluble salts.

Appendix III:
Soil Color

(Quoted and paraphrased from Soil Survey Manual, U.S. Department of Agriculture Handbook No. 18, pp. 189-203.)

Color is the most obvious and easily determined of soil characteristics. Although it has little direct influence on the functioning of the soil, one may infer a great deal about a soil from its color, if it is considered with the other observable features. Thus the significance of soil color is almost entirely an indirect measure of other more important characteristics or qualities that are not so easily and accurately observed. Color is one of the most useful and important characteristics for soil identification, especially when combined with soil structure.

SIGNIFICANCE OF COLOR

The content of organic matter in soil, for example, is a characteristic that is commonly indicated only approximately by soil color. Generally in temperate climates dark-colored soils are relatively higher in organic matter than light-colored soils. In well-drained soils, the colors usually range from very pale brown, through the intermediate browns, to very dark brown or black, as organic matter increases. The most stable part of decomposed organic matter, humus, is darker than the raw or less well-decomposed plant remains. Raw peat is brown, whereas the well-decomposed and more fertile organic soil produced from peat is black or nearly so.

As organic matter is neither all of the same color nor the only coloring matter in soils, soil color by itself is not an exact measure of this important constituent. Well-drained soils of the same relatively high content of organic matter are browner, less nearly black, under high annual temperatures than those of cool regions. Yet the dark clays of warm-temperate and tropical regions, which include some of the blackest soils of the world, seldom contain as much as 3 percent organic matter. Their color may range from medium gray to black with little or no change in the total content of organic matter. Self-mulching black clays with 2 or 3 percent of organic matter may lie side by side with reddish-brown latosolic soils having two or three times as much organic matter. Dark-colored soils low in organic matter may contain compounds of iron and humus, elemental carbon, compounds of manganese, and magnetite. Depth of color depends upon the nature and distribution of the organic matter as well as upon the total amount. In highly alkaline soils, for example, the highly dispersed organic matter apparently coats each soil grain. Such soils are nearly black at relatively low contents of total organic matter.

The red color of soils is generally related to unhydrated iron oxide, although manganese dioxide and partially hydrated iron oxides may also contribute red colors. Since unhydrated iron oxide is relatively unstable under most conditions, red color usually indicates good drainage and good aeration. Strongly red soils are expected on convex surfaces underlain by pervious rocks. Yet many red soils owe their color to inheritance from the parent material and not the soil-forming processes, since the redness in some rocks may persist for centuries even under moist conditions.

In regions where the normal soils have red color, the well-developed red color is one indication that the soils are relatively old or at least the soil material has been subjected to relatively intense weathering for a considerable time. Yet occasionally red colors develop very rapidly. Other things being equal, the red and yellow colors in soil generally increase both in prevalence and in intensity in going from cool regions toward the equator. The intensity of weathering increases with temperature. Then too, many very old land surfaces may be found in warm regions, especially as contrasted with glaciated cool and temperate regions.

The yellow color in soils is also largely due to iron oxides. Yellow colors in the deeper horizons usually indicate a somewhat more moist soil climate than do red colors. The general climatic differences may be in humidity and cloudiness rather than in rainfall. Where associated red and yellow soils are developed from the same kind of parent material, the yellow soils commonly occupy the less convex and more moist sites. Other things being equal, yellow colors are also more common than red colors in regions of high humidity and heavy cloud cover. Apparently, however, it takes a long time to change a yellow soil to a red one, since strongly sloping yellow soils may be found in regions of geologically recent uplift and rapid dissection.

Iron oxides occur in all colors ranging from yellow at the one extreme to red at the other. Thus many brown soils contain relatively large amounts of iron oxides in addition to organic matter.

Well-drained yellowish sands owe their colors to the fact that small amounts of organic matter and other coloring material such as iron oxide are mixed with large amounts of nearly white sand.

Gray and whitish colors of soils are caused by several substances, mainly quartz, kaolin and other clay minerals, carbonates of lime and magnesium, gypsum, various salts, and compounds of ferrous iron. The grayest colors that occur in soils are those of permanently saturated horizons. In these, the iron is in the ferrous form. Some soils are so rich in these compounds as to have nearly pure gray colors that appear bluish.

Imperfectly and poorly drained soils are nearly always mottled with various shades of gray, brown, and yellow, especially within the zone of fluctuation of the water table. In the presence of organic matter, the proportion of gray generally increases with increasing wetness. Wet materials without organic matter rarely have the very light gray color.

A light-gray color may indicate a very low content of organic matter and iron, as in Spodosols, or in sands that consist almost wholly of quartz. Irregular layers of white clay, from which the iron has been removed, are commonly found in the lower parts of poorly drained soils. In arid and semiarid regions, certain soil horizons may be white or nearly white because of the very high content of calcium carbonate, gypsum, or other salts.

Nearly white colors sometimes occur as an inheritance from the parent material, as in Lithosols or Regosols on marls or other white rocks. The failure of soils to accumulate organic matter ordinarily indicates an environment unfavorable to plants and micro-organisms. White soils are almost invariably unproductive naturally, although a few are responsive to good management, such as some weakly developed Rendzinas.

DETERMINATION OF SOIL COLOR

Soil colors are most conveniently measured by comparison with a color chart. The one generally used with soil is a modification of the Munsell color chart and includes only that portion needed for soil colors, about one-fifth of the entire range of color. It consists of some 175 different colored papers, or chips, systematically arranged, according to their Munsell notations, on cards carried in a loose-leaf notebook. The arrangement is by *hue, value,* and *chroma*—the three simple variables that combine to give all colors. *Hue* is the dominant spectral (rainbow) color; it is related to the dominant wavelength of light. *Value* refers to the relative lightness of color and is a function of the total amount of light. *Chroma* (sometimes called saturation) is the relative purity or strength of the spectral color and increases with decreasing grayness.

The nomenclature for soil color consists of two complementary systems: (1) Color names and (2) the Munsell notation of color. Neither of these alone is adequate for all purposes. The color names are employed in all descriptions for publication and for general use. The Munsell notation is used to supplement the color names wherever greater precision is needed, as a convenient abbreviation in field descriptions, for expression of the specific relations between colors, and for statistical treatment of color data. The Munsell notation is especially useful for international correlation, since no translation of color names is needed. The names for soil colors are common terms now so defined as to obtain uniformity and yet accord, as nearly as possible, with past usage by soil scientists.

The Munsell notation for color consists of separate notations for hue, value, and chroma, which are combined in that order to form the color designation. The symbol for hue is the letter abbreviation of the color of the rainbow (**R** for red, **YR** for yellow-red, or orange, **Y** for yellow) preceded by numbers from 0 to 10. Within each letter range, the hue becomes more yellow and less red as the numbers increase. The middle of the letter range is at 5; the zero point coincides with the 10 point of the next redder hue. Thus 5YR is in the middle of the yellow-red hue, which extends from 10R (zero YR) to 10YR (zero Y).

The notation for value consists of numbers from 0, for absolute black, to 10, for absolute white. Thus a color of value 5/ is visually midway between absolute white and absolute black. One of value 6/ is slightly less dark, 60 percent of the way from black to white and midway between values 5/ and 7/.

The notation for chroma consists of numbers beginning at 0 for neutral grays and increasing at equal intervals to a maximum of about 20, which is never really approached in soil. For absolute achromatic colors (pure grays, white, and black), which have zero chroma and no hue, the letter **N** (neutral) takes the place of a hue designation.

In writing the Munsell notation, the order is hue, value, chroma, with a space between the hue letter and the succeeding value number, and a virgule between the two numbers for value and chroma. If expression beyond the whole numbers is desired, decimals are always used, never fractions. Thus the notation for a color of hue 5YR, value 5, chroma 6, is 5YR 5/6, a yellowish-red. Since color determinations cannot be made precisely in the field—generally no closer than half the interval between colors in the chart—expression of color should ordinarily be to the nearest color chip.

In using the color chart, accurate comparison is obtained by holding the soil sample above the color chips being compared. Rarely will the color of the sample be perfectly matched by any color in the chart. It should be evident, however, which colors the sample lies between, and which is the closest match. The principal difficulties encountered in using the soil color chart are (1) in selecting the appropriate hue card, (2) in determining colors that are intermediate between the hues in the chart, and (3) in distinguishing between value and chroma where chromas are strong. In addition, the chart does not include some extreme dark, strong (low value, high chroma) colors occasionally encountered in moist soils.

While important details should be given, long and involved designations of color should generally be avoided, especially with variegated or mottled colors. In these, only the extreme or dominant colors need be stated. Similarly, in giving the color names and Munsell notations for both the dry and moist colors, an abbreviated form, such as "reddish brown (5YR 4/4; 3/4, moist)," simplifies the statement.

Glossary

A horizon. Mineral horizon formed at the soil surface.

Absorbing complex. The materials in the soil that hold water and chemical compounds, mainly on their surfaces. They are chiefly the fine mineral matter and organic matter.

AC soil. A soil having a profile containing only A and C horizons, with no clearly developed B horizon. *See also* **Soil horizon.**

Acid soil. Soil with a pH value < 7.0.

Actinomycetes. A nontaxonomic term applied to a group of organisms with characteristics intermediate between the simple bacteria and the true fungi. Most soil actinomycetes are unicellular microorganisms that produce a slender branched mycelium and sporulate by segmentation of the entire mycelium or, more commonly, by segmentation of special hyphae. Includes many but not all organisms belonging to the order *Actinomycetales*.

Additive. A material added to fertilizer to improve its chemical or physical condition. An additive to liquid fertilizer might prevent crystals from forming in the liquid at temperatures where crystallization would normally take place.

Adsorb. To remove a substance in solution to a solid surface or a separate phase; to accumulate on the surface.

Adsorption. The attachment of compounds or ionic parts of salts to a surface or another phase. Nutrients in solution (ions) carrying a positive charge become attached to (adsorbed by) negatively charged soil particles.

Aerate. To impregnate with a gas, usually air.

Aeration, soil. The process by which air in the soil is replaced by air from the atmosphere. In well-aerated soil, the soil air is very similar in composition to the atmosphere above the soil. Poorly aerated soils usually contain a much higher percentage of carbon dioxide and a correspondingly lower percentage of oxygen than the atmosphere above the soil. The rate of aeration depends largely on the volume and continuity of pores within the soil.

Aeration porosity. *See* **Air porosity.**

Aerobic. (i) Having molecular oxygen as a part of the environment. (ii) Growing only in the presence of molecular oxygen, as aerobic organisms. (iii) Occurring only in the presence of molecular oxygen (said of certain chemical or biochemical processes such as aerobic decomposition).

Aggregate (of soil). Many fine soil particles held in a single mass or cluster, such as a clod, crumb, block, or prism. Many properties of the aggregate differ from those of an equal mass of unaggregated soil.

Agric horizon. A mineral soil horizon in which clay, silt, and humus derived from an overlying cultivated and fertilized layer have accumulated. The wormholes and illuvial clay, silt, and humus occupy at least 5 percent of the horizon by volume. The illuvial clay and humus occur as horizontal lamellae or fibers, or as coatings on ped surfaces or in wormholes.

Agronomy. A specialization of agriculture concerned with the theory and practice of field crop production and soil management. The scientific management of land.

Air porosity. The proportion of the bulk volume of the soil that is filled with air at any given time or under a given condition such as a specified moisture tension.

Air-dry. (i) The state of dryness (of a soil) at equilibrium with the moisture content in the surrounding atmosphere. The actual moisture content will depend upon the relative humidity and the temperature of the surrounding atmosphere. (ii) To allow to reach equilibrium in moisture content with the surrounding atmosphere.

Albic horizon. A mineral soil horizon from which clay and free iron oxides have been removed or in which the oxides have been segregated to the extent that the color of the horizon is determined primarily by the color of the primary sand and silt particles rather than by coatings on these particles.

Albolls. Mollisols that have an albic horizon immediately below the mollic epipedon. These soils have an argillic or natric horizon and mottles, iron-manganese concretions, or both, within the albic, argillic, or natric horizon (a suborder in the USDA soil taxonomy).

Alfisols. Mineral soils that have umbric or ochric epipedons, argillic horizons, and that hold water at less than 15 bars tension during at least three months when the soil is warm enough for plants to grow outdoors. Alfisols have a mean annual soil temperature of less than 8°C or a base saturation in the lower part of the argillic horizon of 35 percent or more when measured at pH 8.2 (an order in the USDA soil taxonomy).

Alkaline soil. Any soil having a pH greater than 7.0.

Alkalinity, soil. The degree of intensity of alkalinity in a soil, expressed by a value greater than 7.0 for the soil pH.

Alluvial soil. (i) A soil developing from recently deposited alluvium and exhibiting essentially no horizon development or modification of the recently deposited materials. (ii) When capitalized the term refers to a Marbut Great Soil Group of the azonal order consisting of soils with little or no modification of the recent sediment in which they are forming (indicated by absence of a B horizon).

Alluvium. Sand, mud, and other sediments deposited on land by streams.

Alumino-silicates. Compounds containing aluminum, silicon, and oxygen atoms as main constituents.

Amendment. (i) An alteration of the properties of a soil, and thereby of the soil, by the addition of substances such as lime, gypsum, or sawdust to the soil for the purpose of making it more suitable for the production of plants. (ii) Any such substance used for this purpose. Strickly speaking, fertilizers constitute a special group of soil amendments.

Amino acids. Nitrogen-containing organic compounds, large numbers of which link together in the formation of a protein molecule. Each amino acid molecule contains one or more amino ($-NH_2$) groups and at least one carboxyl ($-COOH$) group. In addition, some amino acids (cystine and methionine) contain sulfur.

Ammonia. A colorless gas composed of one atom of nitrogen and three atoms of hydrogen. Ammonia liquified under pressure is used as a fertilizer.

Ammonification. The biochemical process whereby ammoniacal nitrogen is released from nitrogen-containing organic compounds.

Ammonium fixation. Adsorption of ammonium ions (NH_4+) by the mineral fraction of the soil in forms that cannot be replaced by a neutral potassium salt solution.

Anaerobic. (i) The absence of molecular oxygen. (ii) Growing in the absence of molecular oxygen (such as anerobic bacteria). (iii) Occurring in the absence of molecular oxygen (as a biochemical process).

Andepts. Inceptisols that have formed either in vitric pyroclastic materials or that have low bulk density and large amounts of amorphous materials, or both. Andepts are not saturated with water long enough to limit their use for most crops (a suborder in the USDA soil taxonomy).

Angular cobbly. *See* **Coarse fragments** and **Cobbly.**

Anhydrous. Dry, or without water. Anhydrous ammonia is water free, in contrast to the water solution of ammonia commonly known as *household ammonia.*

Anion. An ion carrying a negative charge of electricity.

Anion-exchange capacity. The sum total of exchangeable anions that a soil can adsorb. Expressed as milliequivalents per 100 grams of soil (or of other adsorbing material).

Anthropic epipedon. A surface layer of mineral soil that has the same requirements as the mollic epipedon with respect to color, thickness, organic carbon content, consistency, and base saturation, but that has more than 250 parts per million of P_2O_5 soluble in 1 percent citric acid, or is dry more than 10 months (cumulative) during the period when not irrigated. The anthropic epipedon forms under long, continuous cultivation and fertilization.

Anthropic soil. A soil produced from a natural soil or other earthly deposit by human work with new characteristics that make it different from the natural soil. Examples include deep, black surface soils resulting from centuries of manuring, and naturally acid soils that have lost their distinguishing features due to many centuries of liming and use for grass.

Antibiosis. Opposed to living. Antibiotics that suppress some microorganisms.

Antibiotic. A substance produced by one species of organism that, in low concentrations, will kill or inhibit growth of certain other organisms.

Appatite. A native phosphate of lime. The name is given to the chief mineral of phosphate rock and the inorganic compound of bone.

Aqua ammonia. A water solution of ammonia.

Aqualfs. Alfisols that are saturated with water for periods long enough to limit their use for most crops other than pasture or woodland unless they are artificially drained. Aqualfs have mottles, iron-manganese concretions or gray colors immediately below the A1 or Ap horizons, and gray colors in the argillic horizon (a suborder of the USDA soil taxonomy).

Aquents. Entisols that are saturated with water for periods long enough to limit their use for most crops other that pasture unless they are artificially drained. Aquents have low chromas or distinct mottles within 50 centimeters of the surface, or are saturated with water at all times (a suborder of the USDA soil taxonomy).

Aquepts. Inceptisols that are saturated with water for periods long enough to limit their use for most crops other than pasture or woodland unless they are artificially drained. Aquepts have either a histic or umbric epipedon and gray colors within 50 centimeters, or an ochric epipedon underlain by a cambic horizon with gray colors, or have sodium saturation of 15 percent or more (a suborder in the USDA soil taxonomy).

Aquic. A mostly reducing soil moisture regime nearly free of dissolved oxygen due to saturation by ground water or its capillary fringe and occurring at periods when the soil temperature at 50 centimeters is above 5°C.

Aquifer. A water-bearing formation through which water moves more readily than in adjacent formations of lower permeability.

Aquods. Spodosols that are saturated with water for periods long enough to limit their use for most crops other than pasture unless they are artificially drained. Aquods may have a histic epipedon, an albic horizon that is mottled or contains a duripan, or mottling or gray colors within or immediately below the spodic horizon (a suborder of the USDA soil taxonomy).

Aquolls. Mollisols that are saturated with water for periods long enough to limit their use for most crops other than pasture unless they are artificially drained. Aquolls may have a histic epipedon, a sodium saturation in the upper part of the mollic epipedon of more than 15 percent that decreases with depth, or mottles or gray colors within or immediately below the mollic epipedon (a suborder of the USDA soil taxonomy).

Aquox. Oxisols that have continuous plinthite near the surface or that are saturated with water sometime during the year if not artificially drained. Aquox have either a histic epipedon or mottles or colors indicative of poor drainage within the oxic horizon, or both (a suborder of the USDA soil taxonomy).

Aquults. Ultisols that are saturated with water for periods long enough to limit their use for most crops other than pasture or woodland unless they are artificially drained. Aquults have mottles, iron-manganese concretions or gray colors immediately below the A1 or Ap horizons, and gray colors in the argillic horizon (a suborder of the USDA soil taxonomy).

Arents. Entisols that contain recognizable fragments of pedogenic horizons that have been mixed by mechanical disturbance. Arents are not saturated with water for periods long enough to limit their use for most crops (a suborder of the USDA soil taxonomy).

Argids. Aridisols that have an argillic or a natric horizon (a suborder of the USDA soil taxonomy).

Argillic horizon. A mineral soil horizon that is characterized by the illuvial accumulation of layer-lattice silicate clays. The argillic horizon has a certain minimum thickness depending on the thickness of the solum, a minimum quantity of clay in comparison with an overlying eluvial horizon depending on the clay content of the eluvial horizon, and usually has coatings of oriented clay on the surface of pores or peds or bridging sand grains.

Arid climate. A very dry climate like that of desert or semidesert regions where there is only enough water for widely spaced desert plants. The limits of precipitation vary widely according to temperature, with an upper limit for cool regions of less than 254 millimeters and for tropical regions of as much as 508 millimeters. (The precipitation-effectiveness index ranges from 0 to about 16.)

Arid region. Areas where the potential water losses by evaporation and transpiration are greater than the amount of water supplied by precipitation. In the United States this area is broadly considered to be the dry parts of the seventeen western states.

Aridic. A soil moisture regime that has no moisture available for plants for more than half the cumulative time that the soil temperature at a depth of 50 centimeters is above 5°C, and has no period as long as 90 consecutive days when there is moisture for plants while the soil temperature at 50 centimeters is continuously above 8°C.

Aridisols. Mineral soils that have an aridic moisture regime, an ochric epipedon, and other pedogenic horizons but no oxic horizon (an order of the USDA soil taxonomy).

Artificial manure. In Europe the term may denote commercial fertilizers. *See* **Compost.**

Ash. The nonvolatile residue resulting from the complete burning of organic matter. It is commonly composed of oxides of such elements as silicon, aluminum, iron, calcium, magnesium, and potassium.

Assimilation. Conversion of substances taken from the outside into living tissue of plants or animals.

Association, soil. *See* **Soil association.**

Autochthonous flora. That portion of the microflora presumed to subsist on the more resistant soil organic matter and to be little affected by the addition of fresh organic materials. Contrast with zymogenous flora.

Autotrophic. Capable of utilizing carbon dioxide and/or carbonates as a sole or major source of carbon and of obtaining energy for carbon reduction and biosynthetic processes from radiant energy (photoautotroph) or oxidation of inorganic substances (chemoautotroph).

Auxins. Organic substances that cause lengthening of the stem when applied in low concentrations to shoots of growing plants.

Available nutrient. That portion of any element or compound in the soil that can be readily absorbed and assimilated by growing plants. (*Available* should not be confused with *exchangeable*.)

Available water. The portion of water in a soil that can be readily absorbed by plant roots. Considered by most scientists to be that water held in the soil against a pressure of up to approximately 15 bars.

B horizon. A horizon that forms below an A, E, or O horizon.

Badland. A land type generally devoid of vegetation and broken by an intricate maze of narrow ravines, sharp crests, and pinnacles resulting from serious erosion of soft geologic materials. Most common in arid or semiarid regions. A miscellaneous land type.

Banding (of fertilizers). The placement of fertilizers in the soil in continuous narrow ribbons, usually at specified distances from the seeds or plants. The fertilizer bands are covered by the soil but not mixed with it.

Bar. A unit of pressure equal to 1 million dynes per square centimeter.

Base-saturation percentage. The extent to which the adsorption complex of a soil is saturated with exchangeable cations other than hydrogen. It is expressed as a percentage of the total cation-exchange capacity.

Basin irrigation (or level borders). The application of irrigation water to level areas that are surrounded by border ridges or levees. Usually irrigation water is applied at rates greater than the water intake rate of the soil. The water may stand on uncropped soils for several days until the soil is well soaked; then any excess may be used on other fields. The water may stand a few hours on fields having a growing crop.

Basin listing. A method of tillage that creates small basins by damming lister furrows at regular intervals of about 1.3 to 3.5 meters. This method is a modification of ordinary listing and is carried out approximately on the contour on nearly level or gently sloping soils as a means of encouraging water to enter the soil rather than to run off the surface.

BC soil. A soil profile with B and C horizons but with little or no A horizon.

Bedding soil. The arrangement of the surface of fields by plowing and grading into a series of elevated beds separated by shallow ditches for drainage.

Bedrock. The solid rock underlying soils and the regolith in depths ranging from zero (where exposed by erosion) to several hundred meters.

Bench terrace. *See* **Terrace.**

Biological interchange. The interchange of elements between organic and inorganic states in a soil or other substrate through the agency of biological activity. It results from biological decomposition of organic compounds and the liberation of inorganic materials (mineralization), and from the utilization of inorganic materials in the synthesis of microbial tissue (immobilization). Both processes commonly proceed continuously in soils.

Biosequence. A sequence of related soils that differ one from the other, primarily because of differences in kinds and numbers of soil organisms as a soil-forming factor.

Bisect. A profile of plants and a soil showing the vertical and lateral distribution of roots and tops in their natural positions.

Black earth. A term used by some as synonymous with chernozem, by others (in Australia) to describe self-mulching black clays.

Blocky soil structure. *See* **Soil structure.**

Blown-out land. Areas from which all or almost all of the soil and soil material has been removed by wind erosion. Usually barren, shallow depressions with a flat or irregular floor consisting of a more resistant layer and/or an accumulation of pebbles, or a wet zone immediately above a water table. Usually unfit for crop production. A miscellaneous land type.

Blowout. A small area of blown-out land.

Bog iron-ore. Impure ferruginous deposits developed in bogs or swamps by the chemical or biochemical oxidation of iron carried in solution.

Bonds. Chemical forces holding atoms together to form molecules.

Boralfs. Alfisols that have formed in cool places. Boralfs have frigid or cryic but not pergelic temperature regimes, and have udic moisture regimes. Boralfs are not saturated with water for periods long enough to limit their use for most crops (a suborder of the USDA soil taxonomy).

Border strips. *See* **Buffer strips.**

Border-strip irrigation. *See* **Irrigation.**

Borolls. Mollisols with a mean annual soil temperature of less than 8°C and never dry for 60 consecutive days or more within the 3 months following the summer solstice. Borolls do not contain material that has more than 40 percent $CaCO_3$ equivalent unless they have a calcic horizon, and they are not saturated with water for periods long enough to limit their use for most crops (a suborder in the USDA soil taxonomy).

Bottomland. *See* **Floodplain.**

Breccia. A rock composed of coarse angular fragments cemented together.

Broad-base terrace. *See* **Terrace.**

Buffer, buffering. Substances in the soil that act chemically to resist changes in reaction or pH. The buffering action is due mainly to clay and very fine organic matter. Highly weathered tropical clays are less active buffers than most less-weathered silicate clays. Thus, with the same degree of acidity, or pH, more lime is required to neutralize (i) a clayey soil than a sandy soil, (ii) a soil rich in organic matter than one low in organic matter, or (iii) a sandy loam in Michigan, for example, than a sandy loam in central Alabama.

Buffer compounds, soil. The clay, organic matter, and compounds such as carbonates and phosphates that enable the soil to resist appreciable changes in pH.

Buffer strips. Established strips of perennial grass or other erosion-resisting vegetation, usually on the contour in cultivated fields, to reduce runoff and erosion (also referred to as *border strips* or *field border strips*).

Bulk density, soil. The mass of dry soil per unit bulk volume. The bulk volume is determined before drying to constant weight at 105°C.

Bulk specific gravity. The ratio of the bulk density of a soil to the mass of unit volume of water.

Bulk volume. The volume, including the solids and the pores, of an arbitrary soil mass.

Buried soil. Soil covered by an alluvial, loessal, or other deposit, usually to a depth greater than the thickness of the solum.

C horizon. *See* **Soil horizon.**

Calcareous soil. Soil containing sufficient free calcium carbonate or calcium-magnesium carbonate to effervesce visibly when treated with cold 0.1 N hydrochloric acid.

Calcic horizon. A mineral soil horizon of secondary carbonate enrichment that is more than 15 centimeters thick, has a calcium carbonate equivalent of more than 15 percent, and has at least 5 percent more calcium carbonate equivalent than the underlying C horizon.

Caliche. (i) A layer near the surface, more or less cemented by secondary carbonates of calcium or magnesium precipitated from the soil solution. It may occur as a soft, thin soil horizon, as a hard, thick bed just beneath the solum, or as a surface layer exposed by erosion. Not a geologic deposit. (ii) Alluvium cemented with sodium nitrate, chloride, and/or other soluble salts in the nitrate deposits of Chile and Peru.

Cambic horizon. A mineral soil that has a texture of loamy, very fine sand or finer, has soil structure rather than rock structure, contains some weatherable minerals, and is characterized by the alteration or removal of mineral material as indicated by mottling or gray colors, stronger chromas or redder hues than in underlying horizons, or the removal of carbonates. The cambic horizon lacks cementation or induration and has too few evidences of illuviation to meet the requirements of the argillic or spodic horizon.

Capillary fringe. A zone just above the water table (zero gauge pressure) that remains almost saturated. (The extent and the degree of definition of the capillary fringe depend upon the size distribution of pores.)

Carbon-nitrogen (C–N) ratio. The ratio of the weight of organic carbon to the weight of total nitrogen (mineral plus organic forms) in soil or organic material. Often used synonymously with carbon–organic nitrogen ratio when mineral nitrogen levels are low.

Category. Any one of the ranks of the system of soil classification in which soils are grouped on the basis of their characteristics.

Catena. A sequence of soils of about the same age, derived from similar parent material, and occurring under similar climatic conditions, but having different characteristics due to variation in relief and drainage.

Cation exchange. The interchange between a cation in solution and another cation on the surface of any surface-active material such as clay colloid or organic colloid.

Cation-exchange capacity (CEC). The sum total of exchangeable cations that a soil can adsorb. Expressed in milliequivalents per 100 grams or per gram of soil (or other exchangers such as clay).

Cemented. Indurated; having a hard, brittle consistency because the particles are held together by cementing substances such as humus; calcium carbonate; or the oxides of silicon, iron, and aluminum. The hardness and brittleness persist even when wet.

Chroma. The relative purity, strength, or saturation of a color; directly related to the dominance of the determining wavelength of the light and inversely related to grayness; one of the three variables of color.

Chronosequence. A sequence of related soils that differ in certain properties primarily as a result of time as a soil-forming factor.

Class, soil. A group of soils having a definite range in a particular property such as acidity, degree of slope, texture, structure, land-use capability, degree of erosion, or drainage.

Clay. (i) A soil separate consisting of particles less than 2 microns in equivalent diameter. (ii) A textural class.

Clay films. Coatings of clay on the surfaces of soil peds, and mineral grains and in soil pores (also called *clay skins, clay flows, illuviation cutans, argillans,* or *tonhautchen*).

Clay loam. A textural class.

Clay mineral. (i) Naturally occurring inorganic crystalline material found in soils and other earthy deposits, the particles being of clay size, that is, less than 2 microns in diameter. (ii) Material as described under (i) but not limited by particle size.

Clay pan. A dense, compact layer in the subsoil having a much higher clay content than the overlying material, from which it is separated by a sharply defined boundary; formed by downward movement of clay or by synthesis of clay in place during soil formation. Clay pans are usually hard when dry, and plastic and sticky when wet. Also, they usually impede the movement of water and air and the growth of plant roots.

Clayey. Containing large amounts of clay or having properties similar to those of clay.

Climosequence. A sequence of related soils that differ in certain properties primarily as a result of the effect of climate as a soil-forming factor.

Clinosequence. A group of related soils that differ in certain properties primarily as a result of the effect of the degree of slope on which they were formed.

Clod. A compact, coherent mass of soil ranging in size from 0.5 centimeters to as much as 20 or 25 centimeters; produced artificially, usually by human activity such as plowing or digging, especially when these operations are performed on soils that are either too wet or too dry for normal tillage operations.

Coarse fragments. Rock or mineral particles greater than 2 millimeters in diameter.

Cobbly. Containing appreciable quantities of cobblestones (said of soil and land). The term *angular cobbly* is used when the fragments are less rounded.

Colloid. A substance in a state of fine subdivision with particles from 0.00001 to 0.0000001 centimeters (1 micron to 1 millimicron).

Compost. A mixture of various decaying organic substances, such as dead leaves or manure that is used for fertilizing soil.

Concretion. A local concentration of a chemical compound, such as calcium carbonate or iron oxide, in the form of a grain or nodule of varying size, shape, hardness, and color.

Consistency. (i) The resistance of a material to deformation or rupture. (ii) The degree of cohesion or adhesion of the soil mass. Terms used for describing consistency at various soil moisture contents are:

Wet soil. Nonsticky, slightly sticky, sticky, very sticky, nonplastic, slightly plastic, plastic, and very plastic.

Moist soil. Loose, very friable, friable, firm, very firm, and extremely firm.

Dry soil. Loose, soft, slightly hard, hard, very hard, and extremely hard.

Cementation. Weakly cemented, strongly cemented, and indurated.

Crotovina. A former animal burrow in one soil horizon that has been filled with organic matter or material from another horizon (also spelled *krotovina*).

Crumb. A soft, porous, more-or-less-rounded ped from 1 to 5 millimeters in diameter.

Crumb structure. A structural condition in which most of the peds are crumbs.

Crust. A surface layer on soils, ranging in thickness from a few millimeters to perhaps as much as a centimeter, that is much more compact, hard, and brittle when dry than the material immediately beneath it.

Cryic. A soil temperature regime that has mean annual soil temperatures of more than 2.8°C but less than 8°C, more than 5°C difference between mean summer and mean winter soil temperatures at 50 centimeters, and cold summer temperatures.

Crystal structure. The orderly arrangement of atoms in a crystalline material.

Crystalline rock. A rock consisting of various minerals that have crystallized in place from magma.

Cultivation. A tillage operation used in preparing land for seeding or transplanting or later for weed control and for loosening the soil.

Decomposition. The process of resolving into constituent parts. Mineral elements are generally among the terminal products.

Deflation. The removal of fine soil particles from soil by wind.

Deflocculate: (i) To separate the individual components of compound particles by chemical and/or physical means. (ii) To cause the particles of the disperse phase of a colloidal system to become suspended in the dispersion medium.

Denitrification. The biochemical reduction of nitrate or nitrite to gaseous nitrogen either as molecular nitrogen or as an oxide of nitrogen.

Deposit. Material left in a new position by a natural transporting agent such as water, wind, ice, or gravity, or by human activity.

Desert crust. A hard layer, containing calcium carbonate, gypsum, or other binding material, exposed at the surface in desert regions.

Desert pavement. The layer of gravel or stones left on the land surface in desert regions after removal of fine material by wind erosion.

Durinodes. Weakly cemented to indurated soil nodules cemented with SiO_2. Durinodes break down in concentrated KOH after treatment with HCl to remove carbonates, but do not break down on treatment with concentrated HCl alone.

Duripan. A mineral soil horizon that is cemented by silica, usually opal or microcrystalline forms of silica, to the point that air-dry fragments will not slake in water or HCl. A duripan may also have accessory cement such as iron oxide or calcium carbonate.

Edaphic. (i) Of or pertaining to the soil. (ii) Resulting from or influenced by factors inherent in the soil or other substrate, rather than by climatic factors.

Edaphology. The science that deals with the influence of soils on living things, particularly plants, including human use of land for plant growth.

Effective precipitation. The portion of the total precipitation that becomes available for plant growth.

Eluvial horizon. A soil horizon that has been formed by the process of eluviation.

Eluviation. The removal of soil material in suspension (or in solution) from a layer or layers of soil. (Usually, the loss of material in solution is described by the term *leaching*.)

Entisols. Mineral soils that have no distinct pedogenic horizons within 100 centimeters of the soil surface (an order in the USDA soil taxonomy).

Erode. To wear away or remove the land surface by wind, water, or other agents.

Erodible. Susceptible to erosion (expressed by terms such as highly erodible, slightly erodible, and others).

Erosion. (i) The wearing away of the land surface by running water, wind, ice, or other geological agents, including such processes as gravitational creep. (ii) Detachment and movement of soil or rock by water, wind, ice, or gravity.

Evapotranspiration. The combined loss of water from a given area, and during a specified period of time, by evaporation from the soil surface and by transpiration from plants.

Exchange capacity. The total ionic charge of the adsorption complex active in the adsorption of ions.

Family, soil. In soil classification, one of the categories intermediate between the great soil group and the soil series.

Ferrods. Spodosols that have more than six times as much free iron (elemental) than organic carbon in the spodic horizon. Ferrods are rarely saturated with water or do not have characteristics associated with wetness.

Fertility, soil. The status of soil with respect to the amount and availability to plants of elements necessary for plant growth.

Fertilizer. Any organic or inorganic material of natural or synthetic origin that is added to a soil to supply certain elements essential to the growth of plants.

Fibrists. Histosols that have a high content of undecomposed plant fibers and a bulk density less than about 0.1. Fibrists are saturated with water for periods long enough to limit their use for most crops unless they are artificially drained.

Field border strips. *See* **Buffer strips.**

Field capacity (field moisture capacity). The percentage of water remaining in a soil 2 or 3 days after having been saturated and after free drainage has practically ceased. (The percentage may be expressed on the basis of weight or volume.)

Film water. A layer of water surrounding soil particles and varying in thickness from 1 or 2 to perhaps 100 or more molecular layers. Usually considered as that water remaining after drainage has removed free water, because it is not distinguishable in saturated soils.

Fine texture. Consisting of or containing large quantities of the fine fractions, particularly of silt and clay.

Firm. A term describing the consistency of a moist soil that offers distinctly noticeable resistance to crushing but can be crushed with moderate pressure between the thumb and forefinger.

First bottom. The normal floodplain of a stream.

Fixation. The process or processes in a soil by which certain chemical elements essential for plant growth are converted from a soluble or exchangeable form to a much less soluble or a nonexchangeable form, for example, phosphate fixation. Contrast with nitrogen fixation.

Floodplain. A plain, bordering a river, that has been formed from deposits of sediment carried down by the river. When a river rises and overflows its banks, the water spreads over the floodplain; a layer of sediment is deposited at each flood, so that the floodplain gradually rises.

Fluvents. Entisols that form in recent loamy or clayey alluvial deposits, are usually stratified, and have an organic carbon content that decreases irregularly with depth. Fluvents are not saturated with water for periods long enough to limit their use for most crops.

Folists. Histosols that have an accumulation of organic soil materials mainly as forest litter that is less than 100 centimeters deep to rock or to fragmental materials with interstices filled with organic materials. Folists are not saturated with water for periods long enough to limit their use if cropped.

Fragipan. A natural subsurface horizon with high bulk density relative to the solum above, seemingly cemented when dry, but when moist showing a moderate to weak brittleness. The layer is low in organic matter, mottled, slowly or very slowly permeable to water, and usually shows occasional or frequent bleached cracks forming polygons. It may be found in profiles of either cultivated or virgin soils but not in calcareous material.

Friable. A consistency term pertaining to the ease of crumbling of soils.

Frigid. A soil temperature regime that has mean annual soil temperatures of more than 0°C but less than 8°C, more than 5°C difference between mean summer and mean winter soil temperatures at 50 centimeters, and warm summer temperatures. Isofrigid is the same, except the summer and winter temperatures differ by less than 5°C.

Fulvic acid. A term of varied usage but usually referring to the mixture of organic substances remaining in solution upon acidification of a dilute alkali extract from the soil.

Genetic. Resulting from, or produced by, soil-forming processes, for example, a genetic soil profile or a genetic horizon altered from its geologic form.

Gilgai. The microrelief of soils produced by expansion and contraction with changes in moisture. Found in soils that contain large amounts of clay that swell and shrink considerably with wetting and drying. Usually a succession of microbasins and microknolls in nearly level areas or of microvalleys and microridges parallel to the direction of the slope.

Gravelly. Containing appreciable or significant amounts of gravel (used to describe soils or lands).

Gravitational water. Water that moves into, through, or out of the soil under the influence of gravity.

Green manure. Plant material incorporated with the soil while green, or soon after maturity, for improving the soil.

Green manure crop. A crop grown for use as green manure.

Ground water. The portion of the total precipitation that at any particular time is either passing through or standing in the soil and the underlying strata and is free to move under the influence of gravity.

Gypsic horizon. A mineral soil horizon of secondary calcium sulfate enrichment that is more than 15 centimeters thick, has at least 5 percent more gypsum than the C horizon, and in which the product of the thickness in centimeters and the percent calcium sulfate is equal to or greater than 150 percent centimeters.

Hardpan. A hardened soil layer, in the lower A or in the B horizon, caused by cementation of soil particles with organic matter or with materials such as silica, sesquioxides, or calcium carbonate. The hardness does not change appreciably with changes in moisture content, and pieces of the hard layer do not slake in water.

Hemists. Histosols that have an intermediate degree of plant fiber decomposition and a bulk density between about 0.1 and 0.2. Hemists are saturated with water for periods long enough to limit their use for most crops unless they are artificially drained.

Heterotrophic. An organism capable of deriving energy for life processes from the oxidation of organic compounds.

Histic epipedon. A thin organic soil horizon that is saturated with water at some period of the year unless artificially drained and that is at or near the surface of a mineral soil. The histic epipedon has a maximum thickness depending on the kind of materials in the horizon, and the lower limit of organic carbon is the upper limit for the mollic epipedon.

Histosols. Organic soils that have organic soil materials in more than half of the upper 80 centimeters, or that are of any thickness of overlying rock or fragmental materials that have interstices filled with organic soil materials.

Hue. One of the three variables of color. It is caused by light of certain wavelengths and changes with the wavelength.

Humic acid. A mixture of dark-colored substances of indefinite composition extracted from soil with dilute alkali and precipitated by acidification.

Humification. The process involved in the decomposition of organic matter and leading to the formation of humus.

Humods. Spodosols that have accumulated organic carbon and aluminum, but not iron, in the upper part of the spodic horizon. Humods are rarely saturated with water or do not have characteristics associated with wetness.

Humox. Oxisols that are moist all or most of the time and that have a high organic carbon content within the upper meter. Humox have a mean annual soil temperature of less than 22°C and a base saturation within the oxic horizon of less than 35 percent, measured at pH 7.

Humults. Ultisols that have a high content of organic carbon. Humults are not saturated with water for periods long enough to limit their use for most crops.

Humus. That more-or-less-stable fraction of the soil organic matter remaining after the major portion of added plant and animal residues has decomposed. Usually it is dark colored.

Hydrologic cycle. The fate of water from the time of precipitation until the water has been returned to the atmosphere by evapotranspiration and is again ready to be precipitated.

Hyperthermic. A soil temperature regime that has mean annual soil temperatures of 22°C or more and more than 5°C difference between mean summer and mean winter soil temperatures at 50 centimeters. Isohyperthermic is the same except the summer and winter temperatures differ by less than 5°C.

Igneous rock. Rock formed from the cooling and solidification of magma, and that has not been changed appreciably since its formation.

Illite. A hydrous mica.

Illuvial horizon. A soil layer or horizon in which material carried from an overlying layer has been precipitated from solution or deposited from suspension. The layer of accumulation.

Illuviation. The process of decomposition of soil material removed from one horizon to another in the soil, usually from an upper to a lower horizon in the soil profile.

Immature soil. A soil with indistinct or only slightly developed horizons because of the relatively short time it has been subjected to the various soil-forming processes. A soil that has not reached equilibrium with its environment.

Immobilization. The conversion of an element from the inorganic to the organic form in microbial tissues or in plant tissues, thus rendering the element not readily available to the other organisms or the plants.

Impeded drainage. A condition that hinders the movement of water through soils under the influence of gravity.

Impervious. Resistant to penetration by fluids or by roots.

Inceptisols. Mineral soils that have one or more pedogenic horizons in which mineral materials other than carbonates or amorphous silica have been altered or removed but not accumulated to a significant degree. Under certain conditions, Inceptisols may have an ochric, umbric, histic, plaggen, or mollic epipedon. Water is available to the plants more than half of the year or more than three consecutive months during a warm season.

Infiltration. The downward entry of water into soil.

Infiltration rate. A soil characteristic determining or describing the maximum rate at which water can enter soil under specified conditions, including the presence of an excess of water. It has the dimensions of velocity.

Intergrade. A soil that possesses moderately well developed distinguishing characteristics of two or more genetically related great soil groups.

Ions. Atoms, groups of atoms, or compounds that are electrically charged as a result of the loss of electrons (cations) or the gain of electrons (anions).

Iron pan. An indurated soil horizon in which iron oxide is the principal cementing agent.

Irrigation. The artificial application of water to the soil for the benefit of growing crops.

Isohyperthermic. *See* **Hyperthermic.**

Isomesic. *See* **Mesic.**

Isothermic. *See* **Thermic.**

Kaolinite. (i) An alumino-silicate mineral of the 1:1 crystal lattice group, that is, consisting of one silicon tetrahedral layer and one aluminum oxide-hydroxide octahedral layer. (ii) The 1:1 group or family of alumino-silicates.

Krotovina. *See* **Crotovina.**

Lattice. A three-dimensional grid of lines connecting the points representing the centers of atoms or ions in a crystal.

Leaching. The removal of materials in solution from the soil.

Lime, agricultural. A soil amendment consisting principally of calcium carbonate but including magnesium carbonate and perhaps other materials, and used to furnish calcium and magnesium as essential elements for the growth of plants and to neutralize soil acidity.

Lime concretion. An aggregate of precipitated calcium carbonate or other material cemented by precipitated calcium carbonate.

Lime pan. A hardened layer cemented by calcium carbonate.

Lime requirement. The mass of agricultural limestone, or the equivalent of other specified liming material, required per hectare to a soil depth of 15 centimeters to raise the pH of the soil to a desired value under field conditions.

Lithic contact. A boundary between soil and continuous, coherent, underlying material. The underlying material must be sufficiently coherent to make hand digging with a spade impractical. If mineral, it must have a hardness of 3 or more (Mohs scale), and gravel-sized chunks that can be broken out and do not disperse with 15 hours shaking in water or sodium hexameta-phosphate solution.

Loam. A soil textural class.

Loamy. Intermediate in texture, and properties between fine-textured and coarse-textured soils.

Loess. Material transported and deposited by wind and consisting of predominantly silt-sized particles.

Magma. A naturally occurring silicate melt, which may contain suspended silicate crystals or dissolved gases or both.

Marl. Soft and unconsolidated calcium carbonate, usually mixed with varying amounts of clay or other impurities.

Mature soil. A soil with well-developed soil horizons produced by the natural processes of soil formation and essentially in equilibrium with its present environment.

Medium texture. Intermediate between fine-textured and coarse-textured (soils).

Mesic. A soil temperature regime that has mean annual soil temperatures of 8°C or more but less than 15°C, and more than 5°C difference between mean summer and mean winter soil temperatures at 50 centimeters. Isomesic is the same except the summer and winter temperatures differ by less than 5°C.

Mesophilic bacteria. Bacteria whose optimum temperature for growth falls in an intermediate range of approximately 15°C to 45°C.

Metamorphic rock. Rock derived from preexisting rocks but that differ from them in physical, chemical, and mineralogical properties as a result of natural geological processes, principally heat and pressure, originating within the earth. The preexisting rocks may have been igneous, sedimentary, or another form of metamorphic rock.

Microclimate. (i) The climatic condition of a small area resulting from the modification of the general climatic conditions by local differences in elevation or exposure. (ii) The sequence of atmospheric changes within a very small region.

Microfauna. Protozoa and smaller nematodes.

Microflora. Bacteria, including actinomycetes, viruses, and fungi.

Micronutrient. A chemical element necessary in only extremely small amounts (less than 1 part per million in the plant) for the growth of plants. Examples are B, Cl, Cu, Fe, Mn, and Zn. (*Micro* refers to the amount used rather than to its essentiality.)

Microrelief. Small-scale, local differences in topography, including mounds, swales, or pits, that are less than 1 meter in diameter and with elevation differences of up to 2 meters.

Mineral soil. A soil consisting predominantly of, and having its properties determined predominantly by, mineral matter. Usually contains less than 20 percent organic matter, but may contain an organic surface layer up to 30 centimeters thick.

Mineralization. The conversion of an element from an organic form to an inorganic state as a result of microbial decomposition.

Mineralogical analysis. The estimation or determination of the kinds or amounts of minerals present in a rock or in a soil.

Mollic epipedon. A surface horizon of mineral soil that is dark colored and relatively thick, contains at least 0.58 percent organic carbon, is not massive and hard or very hard when dry, has a base saturation of more than 50 percent when measured at pH 7, has less than 250 parts per million of P_2O_5 soluble in 1 percent citric acid, and is dominantly saturated with bivalent cations.

Mollisols. Mineral soils that have a mollic epipedon overlying mineral material with a base saturation of 50 percent or more when measured at pH 7. Mollisols may have an argillic, natric, albic, cambic, gypsic, calcic, or petrocalcic horizon, a histic epipedon, or a duripan, but not an oxic or spodic horizon.

Montmorillonite. An alumino-silicate clay mineral with a 2:1 expanding crystal structure that is, having two silicon tetrahedral layers enclosing an aluminum octahedral layer. Considerable expansion may be caused along the C axis by water moving between silica layers of contiguous units.

Mottled zone. A layer that is marked with spots or blotches of different color or shades of color. The pattern of mottling and the size, abundance, and color contrast of the mottles may vary considerably and should be specified in soil description.

Mottling. Spots or blotches of different color or shades of color interspersed with the dominant color.

Muck. Highly decomposed organic material in which the original plant parts are not recognizable. Contains more mineral matter and is usually darker in color than peat.

Muck soil. (i) A soil containing between 20 and 50 percent of organic matter. (ii) An organic soil in which the organic matter is well decomposed.

Mulch. (i) Any material such as straw, sawdust, leaves, plastic film, loose soil that is spread upon the surface of the soil to protect it and plant roots from the effects of raindrops, soil crusting, freezing, and evaporation. (ii) To apply mulch to the soil surface.

Munsell color system. A color designation system that specifies the relative degrees of the three simple variables of color: hue, value, and chroma. For example: 10YR 6/4 is a color (of soil) with a hue = 10YR, value = 6, and chroma = 4. These notations can be translated into several different systems of color names as desired.

Mycorrhiza. Literally "fungus root." The association, usually symbiotic, or specific fungi with the roots of higher plants.

Natric horizon. A mineral soil horizon that satisfies the requirements of an argillic horizon, but that also has prismatic, columnar, or blocky structure and a subhorizon having more than 15 percent saturation with exchangeable sodium.

Neutral soil. A soil in which the surface layer, at least to normal plow depth, is neither acid nor alkaline in reaction.

Nitrification. Biological oxidation of ammonium to nitrite and nitrate, or a biologically induced increase in the oxidation state of nitrogen.

Nitrogen fixation. Biological conversion of molecular nitrogen (N_2) to organic combinations or to forms utilizable in biological processes.

Ochrepts. Inceptisols formed in cold or temperate climates and that commonly have an ochric epipedon and a cambic horizon. They may have an umbric or mollic epipedon less than 25 centimeters thick or a fragipan or duripan under certain conditions. These soils are not dominated by amorphous materials and are not saturated with water for periods long enough to limit their use for most crops.

Ochric epipedon. A surface horizon of mineral soil that is too light in color, too high in chroma, too low in organic carbon, or too thin to be a plaggen, mollic, umbric, anthropic, or histic epipedon, or that is both hard and massive when dry.

Organic soil. A soil that contains a high percentage of organic matter throughout the solum.

Organic soil materials. Soil materials that are saturated with water and have 17.4 percent or more organic carbon if the mineral fraction is 50 percent or more clay; or 11.6 percent organic carbon if the mineral fraction has no clay, or has proportional intermediate contents, or is never saturated with water, and has 20.3 percent or more organic carbon.

Orthents. Entisols that have either textures of very fine sand or finer in the fine earth fraction, or textures of loamy fine sand or coarser and a coarse fragment content of 35 percent or more and that have an organic carbon content that decreases regularly with depth. Orthents are not saturated with water for periods long enough to limit their use for most crops.

Orthids. Aridisols that have a cambic, calcic, petrocalcic, gypsic, or salic horizon or a duripan, but that lack an argillic or natric horizon.

Orthods. Spodosols that have less than six times as much free iron (elemental) than organic carbon in the spodic horizon but whose ratio of iron to carbon is 0.2 or more. Orthods are not saturated with water for periods long enough to limit their use for most crops.

Orthox. Oxisols that are moist all or most of the time, and that have a low to moderate content of organic carbon within the upper 1 meter or a mean annual soil temperature of 22°C or more.

Ortstein. An indurated layer in the B horizon of spodosols in which the cementing material consists of illuviated sesquioxides (mostly iron) and organic matter.

Oxic horizon. A mineral soil horizon that is at least 30 centimeters thick and that is characterized by the virtual absence of weatherable primary minerals or 2:1 lattice clays, and the presence of 1:1 lattice clays, highly insoluble minerals such as quartz sand, the presence of hydrated oxides of iron and aluminum, the absence of water-dispersible clays, and of low cation-exchange capacity and small amounts of exchangeable bases.

Oxisols. Mineral soils that have an oxic horizon within 2 meters of the surface or plinthite as a continuous phase within 30 centimeters of the surface, and that do not have a spodic or argillic horizon above the oxic horizon.

Paleosols, buried. A soil formed on a landscape during the geologic past and subsequently buried by sedimentation.

Paleosols, exhumed. A formerly buried paleosol that has been exposed on the landscape by the erosive stripping of an overlying mantle of sediment.

Pan. A horizon or layer in soils that is strongly compacted, indurated, or very high in clay content.

Pan, genetic. A natural subsurface soil layer of low or very low permeability, with a high concentration of small particles, and differing in certain physical and chemical properties from the soil immediately above or below the pan.

Pan, pressure or induced. A subsurface horizon or soil layer having a higher bulk density and a lower total porosity than the soil directly above or below it, as a result of pressure that has been applied by normal tillage operations or by other artificial means. Frequently referred to as a plowpan, plow sole, or traffic pan.

Paralithic contact. Similar to a lithic contact except that the mineral material below the contact has a hardness of less than 3 (Mohs scale) and the gravel-sized chunks that can be broken out will partially disperse within 15 hours after shaking in water or sodium hexametaphosphate solution.

Parent material. The unconsolidated and more or less chemically weathered mineral or organic matter from which the solum of soils is developed by pedogenic processes.

Particle density. The mass per unit volume of the soil particles. In technical work, usually expressed as grams per cubic centimeter.

Particle size. The effective diameter of a particle measured by sedimentation, sieving, micrometry, or combinations of these methods.

Particle size analysis. Determination of the various amounts of the different separates in a soil sample, usually by sedimentation, seiving, micrometry, or a combination of these methods.

Peat. Unconsolidated soil material consisting largely of undecomposed, or only slightly decomposed, organic matter accumulated under conditions of excessive moisture.

Peat soil. An organic soil containing more than 50 percent organic matter. Used in the United States to refer to the stage of decomposition of the organic matter. *Peat* refers to the slightly decomposed or undecomposed deposits, and *muck* to the highly decomposed materials.

Ped. A unit of soil structure, such as an aggregate, crumb, prism, block, or granule, formed by natural processes (in contrast with a clod, which is formed artificially).

Pedogenic. *See* **Genetic.**

Pedon. A three-dimensional body of soil with lateral dimensions large enough to permit the study of horizon shapes and relations. Its area ranges from 1 to 10 square meters. Where horizons are intermittent or cyclic and recur at linear intervals of 2 to 7 meters, the pedon includes one-half of the cycle. Where the cycle is less than 2 meters or all horizons are continous and of uniform thickness, the pedon has an area of approximately 1 square meter. If the horizons are cyclic but recur at intervals greater than 7 meters, the pedon reverts to the 1-square-meter size, and more than one soil will usually be represented in each cycle.

Percolation, soil water. The downward movement of water through soil. Especially, the downward flow of water in saturated or nearly saturated soil at hydraulic gradients of the order of 1.0 or less.

Pergelic. A soil temperature regime that has mean annual soil temperatures of less than 0°C. Permafrost is present.

Permafrost. (i) Permanently frozen material underlying the solum. (ii) A perennially frozen soil horizon.

Permeability, soil. (i) The ease with which gases, liquids, or plant roots penetrate or pass through a bulk mass of soil or a layer of soil. Since different soil horizons vary in permeability, the particular horizon under question should be designated. (ii) The property of a porous medium itself that relates to the ease with which gases, liquids, or other substances can pass through it.

Petrocalcic horizon. A continuous, indurated calcic horizon that is cemented by calcium carbonate and, in some places, with magnesium carbonate. It cannot be penetrated with a spade or auger when dry, dry fragments do not slake in water, and it is impenetrable to roots.

Petrogypsic horizon. A continuous, strongly cemented, massive, gypsic horizon that is cemented by calcium sulfate. It can be chipped with a spade when dry. Dry fragments do not slake in water, and it is impenetrable to roots.

pH, soil. The negative logarithm of the hydrogen-ion activity of a soil. The degree of acidity (or alkalinity) of a soil as determined by means of a glass, quinhydrone, or other suitable electrode or by an indicator at a specified moisture content or soil-water ratio, and expressed in terms of the pH scale. (*See also:* **Reaction, soil**)

Phase, soil. A subdivision of a soil type or other unit of classification having characteristics that affect the use and management of the soil but which do not vary sufficiently to differentiate it as a separate type. A variation in a property or characteristic, such as degree of slope, degree of erosion, or content of stones.

Physical weathering. The breakdown of rock and mineral particles into smaller particles by physical forces such as frost action.

Placic horizon. A mineral soil horizon black to dark-reddish and that is usually thin but that may range from 1 to 25 millimeters in thickness. The placic horizon is commonly cemented with iron and is slowly permeable or impenetrable to water and roots.

Plaggen epipedon. A man-made surface horizon more than 50 centimeters thick that is formed by long-continued manuring and mixing.

Plaggepts. Inceptisols that have a plaggen epipedon.

Plastic soil. A soil capable of being molded or deformed continuously and permanently by relatively moderate pressure into various shapes.

Platy. Consisting of soil aggregates that are developed predominantly along the horizontal axis; laminated; flaky.

Plinthite. A nonindurated mixture of iron and aluminum oxides, clay, quartz, and other diluents that commonly occurs as red soil mottles usually arranged in platy, polygonal, or reticulate patterns. Plinthite changes irreversibly to ironstone hardpans or irregular aggregates on exposure to repeated wetting and drying.

Pore space. Total space not occupied by soil particles in a bulk volume of soil.

Pore-size distribution. The volume of the various sizes of pores in a soil. Expressed as percentages of the bulk volume (soil plus soil space).

Porosity. The volume percentage of the total bulk not occupied by solid particles.

Precipitation interception. The stopping, interrupting, and temporary holding of precipitation in any form by a vegetative canopy or vegetation residue.

Primary mineral. A mineral that has not been altered chemically since deposition and crystallization from molten lava.

Prismatic soil structure. A soil structure type with prismlike aggregates that have a vertical axis much longer than the horizontal axes.

Productive soil. A soil in which the chemical, physical, and biological conditions are favorable for the economic production of crops suited to a particular area.

Productivity, soil. The capacity of a soil, in its normal environment, for producing a specified plant or sequence of plants under a specified system of management. The "specified" limitations are necessary since no soil can produce all crops with equal success nor can a single system of management produce the same effect on all soils. Productivity emphasizes the capacity of a soil to produce crops and should be expressed in terms of yields.

Profile, soil. A vertical section of the soil through all its horizons and extending into the parent material.

Psamments. Entisols that have textures of loamy fine sand or coarser in all parts, have less than 35 percent coarse fragments, and are not saturated with water for periods long enough to limit their use for most crops.

Reaction, soil. The degree of acidity or alkalinity of a soil, usually expressed as a pH value. Descriptive terms commonly associated with certain ranges in pH are extremely acid, less than 4.5; very strongly acid, 4.5–5.0; strongly acid, 5.1–5.5; medium acid, 5.6–6.0; slightly acid, 6.1–6.5; neutral, 6.6–7.3; slightly alkaline, 7.4–7.8; moderately alkaline, 7.9–8.4; strongly alkaline, 8.5–9.0; and very strongly alkaline, greater than 9.1.

Regolith. The unconsolidated mantle of weathered rock and soil material on the earth's surface; loose earth materials above solid rock. (Approximately equivalent to the term *soil* as used by many engineers.)

Rendolls. Mollisols that have no argillic or calcic horizon but that contain material with more than 40 percent $CaCO_3$ equivalent within or immediately below the mollic epipedon. Rendolls are not saturated with water for periods long enough to limit their use for most crops.

Residual material. Unconsolidated and partially weathered mineral materials accumulated by disintegration of consolidated rock in place.

Residual soil. A soil formed from, or resting on, consolidated rock of the same kind as that from which it was formed and located in the same place. It is also referred to as *sedentary soil.*

Reticulate mottling. A network of streaks of different color, most commonly found in the deeper profiles of Lateritic soils.

Rhizobia. Bacteria capable of living symbiotically in roots of legumes, from which they receive energy and often utilize molecular nitrogen. Collective common name for the genus *Rhizobium.*

Rhizosphere. The zone of soil where the microbial population is altered both quantitatively and qualitatively by the presence of plant roots.

Rill. A small, intermittent water course with steep sides; usually only a few centimeters deep and, hence, no obstacle to tillage operations.

Runoff. That portion of the precipitation on an area discharged from the area through stream channels. That which is lost without entering the soil is called *surface runoff,* and that which enters the soil before reaching the stream is called *ground-water runoff* or *seepage flow from ground water.* (In soil science, *runoff* usually refers to the water lost by surface flow; in geology and hydraulics, *runoff* usually includes both surface and subsurface flow.)

Salic horizon. A mineral soil horizon of enrichment with secondary salts more soluble in cold water than gypsum. A salic horizon is 15 centimeters or more in thickness and contains at least 2 percent salt; the product of the thickness in centimeters and percent salt by weight is 60 percent centimeter or more.

Sand. (i) A soil particle between 0.05 and 2 millimeters in diameter. (ii) Any one of five soil separates; namely, very coarse sand, coarse sand, medium sand, fine sand, and very fine sand. (iii) A soil textural class.

Sandy. Containing large amounts of sand. (Applied to any one of the soil classes that contains a large percentage of sand.)

Saprists. Histosols that have a high content of plant materials so decomposed that original plant structures cannot be determined, and a bulk density of about 0.2 or more. Saprists are saturated with water for periods long enough to limit their use for most crops unless they are artificially drained.

Saturate. (i) To fill all the voids between soil particles with a liquid. (ii) To form the most concentrated solution possible under a given set of physical conditions in the presence of an excess of the solute. (iii) To fill to capacity, as the adsorption complex with a cation species, for example, H-saturated.

Secondary mineral. A mineral resulting from the decomposition of a primary mineral or from the reprecipitation of the products of decomposition of a primary mineral.

Sedentary soil. *See* **Residual soil.**

Sedimentary rock. A rock formed from materials deposited from suspension or precipitated from solution and usually being more or less consolidated. The principal sedimentary rocks are sandstones, shales, limestones, and conglomerates.

Self-mulching soil. A soil in which the surface layer becomes so well aggregated that it does not crust and seal under the impact of rain, but instead serves as a surface mulch upon drying.

Shaley. (i) Containing a large amount of shale fragments, as a soil. (ii) A soil phase as, for example, shaley phase.

Silica-alumina ratio. The molecules of silicon dioxide (SiO_2) per molecule of aluminum oxide Al_2O_3) in clay minerals or in soils.

Silt. (i) A soil separate consisting of particles between 0.05 and 0.002 millimeters in equivalent diameter. (ii) A soil textural class.

Silt loam. A soil textural class containing a large amount of silt and small quantities of clay and sand.

Silty clay. A soil textural class containing a relatively large amount of silt and clay and a small amount of sand.

Silty clay loam. A soil textural class containing a relatively large amount of silt, a lesser quantity of clay, and a still smaller quantity of sand.

Slickensides. Polished and grooved surfaces produced by one mass sliding past another. Common in Vertisols.

Soil. (i) The unconsolidated mineral material on the immediate surface of the earth that serves as a natural medium for the growth of land plants. (ii) The unconsolidated mineral matter on the surface of the earth that has been subjected to and influenced by genetic and environmental factors of parent material, climate (including moisture and temperature effects), macro- and micro-organisms, and topography, all acting over a period of time and producing a product—soil—that differs from the material from which it is derived in many physical, chemical, biological, and morphological properties and characteristics.

Soil air. The soil atmosphere; the gaseous phase of the soil, being that volume not occupied by solid or liquid.

Soil association. (i) A group of defined and named taxonomic soil units occurring together in an individual and characteristic pattern over a geographic region, comparable to plant associations in many ways. (ii) A mapping unit usually used on general soil maps and sometimes on detailed soil maps in which two or more defined taxonomic units occurring together in a characteristic pattern are mapped as a unit because the scale of the map or the purpose for which it is being made does not require delineation of the individual soils.

Soil complex. A mapping unit used in detailed soil surveys where two or more defined taxonomic units are so intimately intermixed geographically that it is undesirable or impractical, because of the scale being used, to separate them.

Soil conservation. (i) Protection of the soil against physical loss by erosion or against chemical deterioration; that is, excessive loss of fertility by either natural or artificial means. (ii) A combination of all management and land-use methods that safeguard the soil against depletion or deterioration by natural or by man-induced factors. (iii) A division of soil science concerned with soil conservation (i) and (ii).

Soil formation factors. The variable, usually interrelated natural agencies that are active in and responsible for the formation of soil. The factors are usually grouped into five major categories: parent material, climate, organisms, topography, and time.

Soil genesis. (i) The mode of origin of the soil with special reference to the processes or soil-forming factors responsible for the development of the solum, or true soil, from the unconsolidated parent material. (ii) A division of soil science concerned with soil genesis.

Soil geography. A subspecialization of physical geography concerned with the areal distributions of soil types.

Soil horizon. A layer of soil or soil material approximately parallel to the land surface and differing from adjacent genetically related layers in physical, chemical, and biological properties or characteristics such as color, structure, texture, consistency, kinds, and numbers of organisms present, degree of acidity or alkalinity. It is produced by the interaction of soil-forming factors as opposed to strata, which are geologically produced.

Soil management. (i) The sum total of all tillage operations, cropping practices, fertilizers, lime, and other treatments conducted on or applied to a soil for the production of plants. (ii) A division of soil science concerned with the items listed under (i).

Soil mineral. (i) Any mineral that occurs as a part of or in the soil. (ii) A natural inorganic compound with definite physical, chemical, and crystalline properties (within the limits of isomorphism) that occurs in the soil.

Soil mineralogy. A subspecialization of soil science concerned with the homogeneous inorganic materials found in the earth's crust to the depth of weathering or of sedimentation.

Soil moisture. Water contained in the soil.

Soil organic matter. The organic fraction of the soil; includes plant and animal residues at various stages of decomposition, cells and tissues of soil organisms, and substances synthesized by the soil population. Usually determined on soils which have been sieved through a 2 millimeter sieve.

Soil pores. That part of the bulk volume of soil not occupied by soil particles; interstices; voids.

Soil science. That science dealing with soils as a natural resource on the surface of the earth, including soil formation; classification; mapping; the physical, chemical, biological, and fertility properties of soils per se; and these properties in relation to their management for crop production.

Soil separates. Mineral particles less than 2 millimeters in equivalent diameter, ranging between specified size limits. The names and size limits of separates recognized in the United States are very coarse sand, 1 to 2 millimeters; coarse sand, 0.5 to 1 millimeters; medium sand, 0.5 to 0.25 millimeters; fine sand, 0.25 to 0.10 millimeters; very fine sand, 0.10 to 0.05 millimeters; silt, 0.05 to 0.002 millimeters; and clay, less than 0.002 millimeters.

Soil series. The basic unit of soil classification being a subdivision of a family and consisting of soils that are essentially alike in all major profile characteristics except the texture of the A horizon.

Soil solution. The aqueous liquid phase of the soil and its solutes consisting of ions dissociated from the surfaces of the soil particles and of other soluble materials.

Soil structure. The combination or arrangement of primary soil particles into secondary particles, units, or peds. These secondary units may be, but usually are not, arranged in the profile in such a manner as to give a distinctive characteristic pattern. The secondary units are characterized and classified on the basis of size, shape, and degree of distinctness into classes, types, and grades, respectively.

Soil structure classes. A grouping of soil structural units or peds on the basis of size.

Soil structure grades. A grouping or classification of soil structure on the basis of inter- and intra-aggregate adhesion, cohesion, or stability within the profile. Four grades of structure designated from 0 to 3 are recognized as follows:

0. *Structureless.* No observable aggregation or no definite and orderly arrangement of natural lines of weakness. Massive, if coherent; single grain, if not coherent.

1. *Weak.* Poorly formed, indistinct peds, barely observable in place.

2. *Moderate.* Well-formed, distinct peds, moderately durable and evident, but not distinct in undisturbed soil.

3. *Strong.* Durable peds that are quite evident in undisturbed soil, adhere weakly to one another, withstand displacement, and become separated when the soil is disturbed.

Soil survey. The systematic examination, description, classification, and mapping of soils in an area. Soil surveys are classified according to the kind and intensity of field examination.

Soil texture. The relative proportions of the various soil separates in a soil as described by the classes of soil texture. The textural classes may be modified by the addition of suitable adjectives when coarse fragments are present in substantial amounts, for example, "stony silt loam" or "silt loam, stony phase." The sand, loamy sand, and sandy loam are further subdivided on the basis of the proportions of the various sand separates present. The limits of the various classes and subclasses are as follows:

Sand. Soil material that contains 85 percent or more of sand; percentage of silt, plus 1.5 times the percentage of clay, does not exceed 15.

Coarse sand. 25 percent or more very coarse and coarse sand, and less than 50 percent any other one grade of sand.

Sand. 25 percent or more very coarse, coarse, and medium sand, and less than 50 percent very fine sand.

Fine sand. 50 percent or more fine sand (or) less than 25 percent very coarse, coarse, and medium sand, and less than 50 percent very fine sand.

Very fine sand. 50 percent or more very fine sand.

Loamy sand. Soil material that contains at the upper limit 85 to 90 percent sand, the percentage of silt plus 1.5 times the percentage of clay is not less than 70 to 85 percent sand, and the percentage of silt plus twice the percentage of clay does not exceed 30.

Loamy coarse sand. 25 percent or more very coarse and coarse sand, and less than 50 percent any other one grade of sand.

Loamy sand. 25 percent or more very coarse, coarse, and medium sand, and less than 50 percent fine or very fine sand.

Loamy fine sand. 50 percent or more fine sand (or) less than 25 percent very coarse, coarse, and medium sand, and less than 50 percent very fine sand.

Loamy very fine sand. 50 percent or more very fine sand.

Sandy loam. Soil material that contains either 20 percent clay or less, the percentage of silt plus twice the percentage of clay exceeds 30, and 52 percent or more sand; or less than 7 percent clay, less than 50 percent silt, and between 43 percent and 52 percent sand.

Coarse sandy loam. 25 percent or more very coarse and coarse sand, and less than 50 percent of any other one grade of sand.

Sandy loam. 30 percent or more very coarse, coarse, and medium sand, but less than 25 percent very coarse sand, and less than 30 percent very fine or fine sand.

Fine sandy loam. 30 percent or more fine sand and less than 30 percent very fine sand (or) between 15 and 30 percent very coarse, coarse and medium sand.

Very fine sandy loam. 30 percent or more very fine (or) greater than 40 percent fine and very fine sand, at least half of which is very fine sand, and less than 15 percent very coarse, coarse, and medium sand.

Loam. Soil material that contains 7 to 27 percent clay, 28 to 50 percent silt, and less than 52 percent sand.

Silt loam. Soil material that contains 50 percent or more silt and 12 to 27 percent clay (or) 50 to 80 percent silt and less than 12 percent clay.

Silt. Soil material that contains 80 percent or more silt and less than 12 percent clay.

Sandy clay loam. Soil material that contains 20 to 35 percent clay, less than 28 percent silt, and 45 percent or more sand.

Clay loam. Soil material that contains 27 to 40 percent clay and 20 to 45 percent sand.

Silty clay loam. Soil material that contains 27 to 40 percent clay and less than 20 percent sand.

Sandy clay. Soil material that contains 35 percent or more clay and 45 percent or more sand.

Silty clay. Soil material that contains 40 percent or more clay and 40 percent or more silt.

Clay. Soil material that contains 40 percent or more clay, less than 45 percent sand, and less than 40 percent silt.

Solum (plural: *sola*). The upper and most weathered part of the soil profile; the A, E, and B horizons.

Sombric horizon. A subsurface mineral horizon that is darker in color than the overlying horizon but that lacks the properties of a spodic horizon. Common in cool, moist soils of high altitude in tropical regions.

Spodic horizon. A mineral soil horizon that is characterized by the illuvial accumulation of amorphous materials composed of aluminum and organic carbon with or without iron. The spodic horizon has a certain minimum thickness, and a minimum quantity of extractable carbon plus iron plus aluminum in relation to its content of clay.

Spodosols. Mineral soils that have a spodic horizon or a placic horizon that overlies a fragipan.

Stony. Containing sufficient stones to interfere with or to prevent tillage. To be classified as stony, more than 0.01 percent of the soil surface must be covered with stones. Used to modify soil class, such as "stony clay loam" or "clay loam, stony phase."

Stratified. Arranged in or composed of strata or layers.

Strip cropping. The practice of growing crops that requires different types of tillage, such as row and sod, in alternate strips along contours or across the prevailing direction of wind for the purpose of reducing erosion.

Stubble mulch. The stubble of crops or crop residues left essentially in place on the land as a surface cover before and during the preparation of the seedbed and at least partly during the growing of a succeeding crop.

Subsoiling. Breaking of compact subsoils, without inverting them, with a special knifelike instrument (chisel) that is pulled through the soil at depths usually of 30 to 60 centimeters and at spacings usually of 60 to 150 centimeters.

Subsurface tillage. Tillage with a special sweeplike plow or blade that is drawn beneath the surface and cuts plant roots and loosens the soil without inverting it or without incorporating the surface cover.

Terrace. A nearly level strip of land with a more or less abrupt descent along the margin of the sea, a lake, or a river. In soil conservation, a more or less level or horizontal strip of earth usually constructed on a contour designed to reduce erosion. Also called **Bench Terrace.**

Thermic. A soil temperature regime that has mean annual soil temperatures of 15°C or more but less than 22°C, and more than 5°C difference between mean summer and mean winter soil temperatures at 50 centimeters. Isothermic is the same except the summer and winter temperatures differ by less than 5°C.

Tight soil. A compact, relatively impervious and tenacious soil (or subsoil) that may or may not be plastic.

Tile drain. Concrete, ceramic, or plastic pipe placed at suitable depths and spacings in the soil or subsoil to provide water outlets from the soil.

Till. (i) Unstratified glacial drift deposited directly by the ice and consisting of clay, sand, gravel, and boulders intermingled in any proportion. (ii) To plow and prepare for seeding; to seed or cultivate the soil.

Tilth. The physical condition of soil as related to its ease of tillage, fitness as a seedbed, and its impedance to seedling emergence and root penetration.

Toposequence. A sequence of related soils that differ primarily because of topography as a soil formation factor.

Topsoil. (i) The layer of soil moved in cultivation. (ii) The A horizon. (iii) The A1 horizon. (iv) Presumably fertile soil material used to topdress road banks, gardens, and lawns.

Torrerts. Vertisols of arid regions with wide, deep cracks that remain open throughout the year in most years.

Torric. A soil moisture regime defined like aridic moisture regime, but used in a different category of the soil taxonomy.

Torrox. Oxisols that have a torric soil moisture regime.

Transported soil. Any soil that was formed on unconsolidated sedimentary rocks.

Tropepts. Inceptisols that have a mean annual soil temperature of 8°C or more, and less than 5°C difference between mean summer and mean winter temperatures at a depth of 50 centimeters below the surface. Tropepts may have an ochric epipedon and a cambic horizon, or an umbric epipedon, or a mollic epipedon under certain conditions, but no plaggen epipedon, and are not saturated with water for periods long enough to limit their use for most crops.

Udalfs. Alfisols that have a udic soil moisture regime and mesic or warmer soil temperature regimes. Udalfs generally have brownish colors throughout and are not saturated with water for periods long enough to limit their use for most crops.

Uderts. Vertisols of relatively humid regions that have wide, deep cracks that usually remain open continuously for less than 2 months or intermittently for periods that total less than 3 months.

Udic. A soil moisture regime that is neither dry for as long as 90 cumulative days nor for as long as 60 consecutive days in the 90 days following the summer solstice at periods when the soil temperature at 50 centimeters is above 5°C.

Udolls. Mollisols that have a udic soil moisture regime with mean annual soil temperatures of 8°C or more. Udolls have no calcic or gypsic horizon and are not saturated with water for periods long enough to limit their use for most crops.

Udults. Ultisols that have low to moderate amount of organic carbon, reddish or yellowish argillic horizons, and a udic soil moisture regime. Udults are not saturated with water for periods long enough to limit their use for most crops.

Ultisols. Mineral soils that have a mineral horizon with a base saturation of less than 35 percent. Ultisols have a mean annual soil temperature of 8°C or higher.

Umbrepts. Inceptisols formed in cold or temperate climates that commonly have an umbric epipedon, but they may have a mollic or an anthropic epipedon 25 millimeters or more thick under certain conditions. These soils are not dominated by amorphous materials and are not saturated with water for periods long enough to limit their use for most crops.

Umbric epipedon. A surface layer of mineral soil that has the same requirements as the mollic epipedon with respect to color, thickness, organic carbon content, consistency, structure, and P_2O_5 content, but that has a base saturation of less than 50 percent when measured at pH 7.

Ustalfs. Alfisols that have an ustic soil moisture regime and mesic or warmer soil temperature regime. Ustalfs are brownish or reddish throughout and are not saturated with water for periods long enough to limit their use for most crops.

Usterts. Vertisols of temperate or tropical regions with wide, deep cracks that usually remain open for periods that total more than 3 months but do not remain open continuously throughout the year, and that have either a mean annual soil temperature of 22°C or more or a mean summer soil temperature and a mean winter soil temperature at 50 centimeters that differ by less than 5°C or have cracks that open and close more than once during the year.

Ustic. A soil moisture regime that is intermediate between the aridic and udic regimes and commonly in temperate subhumid or semiarid regions, or in tropical and subtropical regions with a monsoon climate. A limited amount of moisture is available for plants but occurs at times when the soil temperature is optimum for plant growth.

Ustolls. Mollisols that have an ustic soil moisture regime and mesic or warmer soil temperature regimes. Ustolls may have a calcic, petrocalcic, or gypsic horizon, and are not saturated with water for periods long enough to limit their use for most crops.

Ustox. Oxisols that have an ustic moisture regime and either hyperthermic or isohyperthermic soil temperature regimes or have less than 20 kilograms of organic carbon in the surface cubic meter.

Ustults. Ultisols that have low or moderate amounts of organic carbon, are brownish or reddish throughout, and have an ustic soil moisture regime.

Value, color. The relative lightness or intensity of color and approximately a function of the square root of the total amount of light. One of the three variables of color. (*See also* Munsell color system and Appendix III.)

Vertisols. Mineral soils that have 30 percent or more clay; deep, wide cracks when dry; and either gilgai microrelief, intersecting slickensides, or wedge-shaped structural aggregates tilted at an angle from the horizontal.

Water table. The upper surface of ground water or that level below which the soil is saturated with water; locus of points in soil water at which the hydraulic pressure is equal to atmospheric pressure.

Water table, perched. The water table of a saturated layer of soil that is separated from an underlying saturated layer by an unsaturated layer.

Weathering. All physical and chemical changes produced in rocks, at or near the earth's surface, by atmospheric agents.

Wilting point. Same as "permanent wilting percentage," as defined in standard plant physiology texts.

Windbreak. A planting of trees, shrubs, or other vegetation, usually perpendicular or nearly so, to the principal wind direction. Situated to protect soil, crops, homesteads, and roads against such effects of winds as wind erosion and the drifting of soil and snow.

Xeralfs. Alfisols that have a xeric soil moisture regime. Xeralfs are brownish or reddish throughout.

Xererts. Vertisols of Mediterranean climates with wide, deep cracks that open and close once each year and usually remain open continuously for more than 2 months. Xererts have mean annual soil temperature of less than 22°C.

Xeric. A soil moisture regime common to Mediterranean climates that have moist cool winters and warm dry summers. A limited amount of moisture is present but does not occur at optimum periods for plant growth. Irrigation or summer fallow is commonly necessary for crop production.

Xerolls. Mollisols that have a xeric soil moisture regime. Xerolls may have a calcic, petrocalcic, or gypsic horizon, or a duripan.

Xerults. Ultisols that have low or moderate amounts of organic carbon, are brownish or reddish throughout, and have a xeric soil moisture regime.

Yield, sustained. A continual annual, or periodic, yield of plants or plant material from an area; implies management practices that will maintain the productive capacity of the land.

Bibliography

Abuajameih, M. M. "The Structure of the Pantano Beds in the Northern Tucson Basin." Master's thesis. University of Arizona, 1966.

Alexander, L. T., and J. G. Cady. "Genesis and Hardening of Laterite in Soils." U.S.D.A. *Technical Bulletin* #1281. Washington, D.C.: Soil Conservation Service, U.S. Department of Agriculture, 1962.

Allison, Franklin E. "Nitrogen and Soil Fertility." *The 1957 Yearbook of Agriculture: Soil.* Washington, D.C.: U.S. Government Printing Office, 1957. Pp. 85–94.

Aubert, G. "Soils with Ferruginous or Ferralitic Crusts of Tropical Regions." *Soil Science* 95, no. 4 (1963): 235–42.

Baldwin, Mark; Charles E. Kellogg, and James Thorp. "Soil Classification." *The 1938 Yearbook of Agriculture: Soils and Man.* Washington, D.C.: U.S. Government Printing Office, 1938. Pp. 979–1001.

Barnes, C. P. "Environment of Natural Grassland." *The 1948 Yearbook of Agriculture: Grass.* Washington, D.C.: U.S. Government Printing Office, 1948. Pp. 45–49.

Barry, R. G., and R. J. Chorley. *Atmosphere, Weather, and Climate.* New York: Holt, Reinhart and Winston, 1970.

Basila, M. R. "Hydrogen Bonding Interaction Between Adsorbate Molecules and Surface Hydroxyl Groups on Silica." *Chemical Physics* 35 (1961): 1151–58.

Birkeland, P. W. *Soils and Geomorphology.* New York: Oxford University Press, 1984.

Blake, W. P. "The Caliche of Southern Arizona." *American Institute of Mining and Engineering* 31 (1902): 220–26.

Blank, H. R., and E. W. Tynes. "Formation of Caliche in Situ." *Geological Society of America Bulletin* 75 (1965): 1387–91.

Breazeale, J. F., and H. V. Smith. "Caliche in Arizona." *University of Arizona Agricultural Experimental Station Bulletin* 131 (1930): 419–41.

Bretz, J. H., and L. Horburg. "Caliche in Southeastern New Mexico." *Journal of Geology* 57 (1949): 491–511.

Bridges, E. M., and D. A. Davidson. *Principles and Applications of Soil Geography.* London: Longman, 1982.

Brown, C. H. "The Origin of Caliche on the Northeastern Llano Estacada, Texas." *Journal of Geology* 64 (1956): 1–15.

Brown, L.R. "World Food Resources and Population: The Narrowing Margin." *Population Reference Bulletin* 36, no. 3 (1981).

Buckman, Harry O., and Nyle C. Brady. *The Nature and Properties of Soils.* New York: Macmillan Co., 1971.

Bunting, B. T., *The Geography of Soil.* Chicago: Aldine Publishing Co., 1967.

Buol, S. W. *Soils of Arizona.* Tucson: Agricultural Experiment Station, University of Arizona, 1966.

Buol, S. W., F. D. Hole, and R. J. McCracken. *Soil Genesis and Classification,* 2nd ed. Ames: Iowa State University Press, 1980.

Chang, J. "The Agricultural Potential of the Humid Tropics." *The Geographical Review* 58, no. 3 (1968): 333–61.

Clark, Francis E. "Living Organisms in the Soil." *The 1957 Yearbook of Agriculture: Soil.* Washington, D.C.: U.S. Government Printing Office, 1957. Pp. 157–60.

Colwell, W. E. "Tobacco." *The 1957 Yearbook of Agriculture: Soil.* Washington, D.C.: U.S. Government Printing Office, 1957. Pp. 655–58.

Cooley, D. B. "Geological Environment and Engineering Properties of Caliche in the Tucson Area." Master's thesis. University of Arizona, 1966.

Cooley, M. E. "Description and Origin of Caliche in the Glen-San Juan Canyon Region, Utah and Arizona." *Arizona Geological Society Digest* 4 (1961): 35–41.

Daniels, R. B., and H. J. Kleiss, S. W. Buol, H. J. Byrd, and J. A. Phillips. "Soil Systems in North Carolina." *Bulletin 467.* Raleigh: North Carolina Agricultural Research Service, 1984.

Donahue, R. L., J. C. Shickluna, and L. S. Robertson. *Soils: An Introduction to Soils and Plant Growth,* 3rd ed. Englewood Cliffs, N.J.: Prentice-Hall, 1971.

Eeckman, J. P., and H. Laudelett. "Chemical Stability of Hydrogen Montmorillonite." *Kollodzschr* 178, no. 2 (1961): 99–107.

Eyre, S. R. *Vegetation and Soils.* Chicago: Aldine Publishing Company, 1971.

Foth, H. D. *Fundamentals of Soil Science,* 7th ed. New York: John Wiley and Sons, 1984.

Foth, H. D., and L. M. Turk. *Fundamentals of Soil Science,* 5th ed. New York: John Wiley and Sons, 1972.

Freckman, Diane W., ed. *Nematodes in Soil Ecosystems.* Austin: University of Texas Press, 1982.

Fredrickson, A. F. "Mechanism of Weathering." *Bulletin of the Geological Society of America* 62 (1951): 221–32.

Fripiat, J. J., and A. J. Herbillion. "Formation and Transformation of Clay Minerals in Tropical Soils." *Soils and Tropical Weathering.* Paris: UNESCO, 1971. Pp. 15–22.

Fuller, W. H., and H. E. Ray. "Gypsum and Sulfur-Bearing Amendments." *Agricultural Experiment Station Bulletin* A-27. Tucson: University of Arizona, 1963. Pp. 3–11.

Gersmehl, Philip J. "Soil Taxonomy and Mapping." *Annals of the Association of American Geographers* 67. September 1977: 419–28.

Gibbings, P. N. "The Effects of Caliche on the Strength of Concrete." Master's thesis. University of Arizona, 1932.

Gile, L. H. "A Classification of ca Horizons in Soils of a Desert Region, Dona Ana County, New Mexico." *Soil Science Society of America Proceedings* 25 (1961): 52–61.

_____. "Morphological and Genetic Sequences of Carbonate Accumulation in Desert Soils." *Soil Science* 101 (1966): 347–60.

Greenland, D. J., and P. H. Nye. *The Soil Under Shifting Cultivation.* Bucks, England: Commonwealth Agricultural Bureau, 1965.

Grissom, Perrin H. "The Mississippi: Delta Region." *The 1957 Yearbook of Agriculture: Soil.* Washington, D.C.: U.S. Government Printing Office, 1957. Pp. 524–30.

Hallsworth, E. G., and A. B. Costin. "Studies in Pedogenesis in New South Wales, IV, 'The Ironstone Soils'." *Soil Science* 4 (1953): 24–47.

Hausenbuiller, R. L. *Soil Science: Principles and Practices.* Dubuque, Ia.: William C. Brown Publishers, 1985.

Hendricks, D. M., and Y. H. Havens. *Desert Soils Tour Guide.* Tucson: Soil Science Society of America. August 27, 1970.

Hillel, D. *Applications of Soil Physics.* New York: Academic Press, 1980.

Hole, Francis D., and Gerald A. Neilson. "Soil Genesis Under Prairie." *Proceedings of the Symposium on Prairie and Prairie Restoration.* Galesburg, Ill.: Knox College, 1963.

Hunt, Charles B. *Geology of Soils: Their Evolution, Classification, and Uses.* San Francisco: W. H. Freeman and Company, 1972.

Isnail, F. T. "Biotite Weathering and Clay Formation in Arid and Humid Regions, California." *Soil Science* 109 (1969): 257–61.

Jenny, H., and C. D. Leonard. "Functional Relationships Between Soil Properties and Rainfall." *Soil Science* 38 (1934): 363–81.

Keller, W. D. "Process of Origin and Alteration of Clay Minerals." In *Soil Clay Mineralogy,* eds. C. I. Rich and G. W. Kunze. Chapel Hill: University of North Carolina Press, 1964.

Kellogg, Charles E. "Why a New System of Soil Classification?" *Soil Science* 95, no. 1 (July 1963): 1–5.

Kilmer, V. J., ed. *Handbook of Soils and Climate in Agriculture.* Boca Raton, Fla.: CRC Press, 1982.

Martin, W. P., and Joel E. Fletcher. "Vertical Zonation of Great Soil Groups on Mt. Graham, Arizona, as Correlated with Climate, Vegetation, and Profile Characteristics." *Technical Bulletin* 99. Tucson: University of Arizona, 1943. Pp. 144–47.

Mohrand, E. C. J., and F. A. Van Baren. *Tropical Soils.* London: Interscience Publishers, Ltd., 1954.

Muckernhirn, R. J., and K. C. Berger. "The Northern Lake States." *The 1957 Yearbook of Agriculture: Soil.* Washington, D.C.: U.S. Government Printing Office, 1957. Pp. 547–53.

National Cooperative Soil Survey. *Soil Taxonomy.* Soil Conservation Service. Washington, D.C.: U.S. Government Printing Office, 1975.

Nye, P. H. "Some Soil-Forming Processes in the Humid Tropics, I-IV." *Journal of Soil Science* 5, no. 1 (1954): 7–83.

Olson, G. W. *Field Guide to Soils and the Environment: Applications of Soil Surveys.* New York: Chapman and Hall, 1984.

Pearson, R. W., and L.E. Ensminger. "Southeastern Uplands." *The 1957 Yearbook of Agriculture: Soil* Washington, D.C.: U.S. Government Printing Office, 1957. Pp. 580–95.

Pierre, W. H., and F. F. Riecken. "The Midland Feed Region." *The 1957 Yearbook of Agriculture: Soil.* Washington, D.C.: U.S. Government Printing Office, 1957. Pp. 535–47.

Post, J. L. "Strength Characteristics of Caliche Soils of the Tucson Area." Master's thesis. University of Arizona, 1966.

Prescott, J. A. "The Early Use of the Term Laterite." *Journal of Soil Science* 5, no. 1 (1954): 1–5.

Prescott, J. A., and R. L. Pendleton. *Laterite and Lateritic Soils.* Bucks, England: Commonwealth Agricultural Bureau, 1952.

Price, W. A. "Reynosa Problem of South Texas, and Origin of Caliche." *American Association of Petroleum Geologists Bulletin* 17 (1933): 448–522.

Rao, T. Shesagiri. "Pedogenesis of Some Major Soil Groups in Mysore State, India." *Soils and Tropical Weathering: Proceedings of the Bandung Symposium.* Paris: NUESCO, 1971. Pp. 72–84.

Reeves, C. C., Jr. "Caliche." *Encyclopedia of Earth Sciences,* vol. 6. New York: Holt, Rinehart and Winston, 1970.

————. "Origin, Classification, and Geologic History of Caliche on the Southern High Plains, Texas and Eastern New Mexico." *Journal of Geology* 78 (1970): 352–62.

Rienfenburg, A., and S. J. Buckwold. "The Release of Silica from Soils by the Orthophosphate Anion." *Journal of Soil Science* 5, no. 1 (1954): 106–15.

Rojan, S. V. Govinda, and N. R. Datta Biswas. "Development of Certain Soils in the Subtropical Humid Zone in Southeastern Parts of India, Genesis and Classification of Soils of Machkund Basin." *Soils and Tropical Weathering.* Paris: UNESCO, 1971. Pp. 77–85.

Schnitzer, M., and S. U. Kahn, eds. *Soil Organic Matter.* Amsterdam: Elsevier Scientific Publishing Company, 1978.

Schuylenborgh, J. Von. "Investigations of the Classification and Genesis of Soils Derived from Andesite Tuffs Under Humid Tropical Conditions." *Netherlands Journal of Agricultural Science* 5 (1957): 195–210.

Segalen, P. "Metallic Oxides and Hydroxides in Soils of the Warm and Humid Areas of the World: Formation, Identification, Evolution." *Soils and Tropical Weathering.* Paris: UNESCO, 1971. Pp. 25–37.

Shreve, F., and T. D. Mallery. "The Relation of Caliche to Desert Plants." *Soil Science* 35 (1933): 99–113.

Sigalove, J. J. "Carbon 14 Content and Origin of Caliche." Master's thesis. University of Arizona, 1969.

Simonson, R. W. "Morphology and Classification of the Rugur Soils of India." *Journal of Soil Science* 5, no. 1 (1954): 275–88.

Singer, Michael J., and Donald N. Munns. *Soils: An Introduction.* New York: Macmillan Publishing Company, 1987.

Smith, G. E. "The Physiography of Arizona Valleys and Occurrence of Ground Water." *University of Arizona Agricultural Experiment Station Technical Bulletin,* 77 (1983).

Smith, Guy D. "Lectures on Soil Classification." *Pedologie,* vol. 4. Ghent, Belgium: State University of Ghent, 1965.

Soil Conservation Service. *America's Soil and Water: Condition and Trends.* Washington, D.C.: U.S. Government Printing Office, 1980.

————. "Land Resource Regions and Major Land Resource Areas of the United States." *Agricultural Handbook #296.* Washington, D.C.: U.S. Government Printing Office, 1981.

Soil Science Society of America. *Glossary of Soil Science Terms.* Madison, Wis.: Soil Science Society of America, 1975.

Soil Survey Staff. *Soil Survey Manual, Agricultural Handbook #18.* Washington, D.C.: U.S. Government Printing Office, 1951.

————. *Supplement to Soil Classification, A Comprehensive System, 7th Approximation.* Washington, D.C.: Soil Conservation Service, U.S. Department of Agriculture, 1964.

Steila, Donald, *The Geography of Soils.* Englewood Cliffs, N.J.: Prentice-Hall, 1976.

Steila, D., and T. G. Roehrig. *Soil Science Investigations.* Charlotte: University of North Carolina, 1984.

Stobbe, P. C., and J. R. Wright. "Modern Concepts of the Genesis of Podzols." *Soil Science Society of America Proceedings* 23 (1959): 161–63.

Stuart, D. M., M. A. Fosburg, and G.C. Lewis. "Caliche in Southwestern Idaho." *Soil Science Society of America Proceedings* 25 (1961): 132–35.

Syliss, E., G. K. Rennie, C. Smart, and B. A. Pethica. "Anomalous Water." *Nature* 222 (1969): 159–61.

Thorne, D. W., and L. F. Seatz. "Acid, Alkaline, Alkali, and Saline Soil." In *Chemistry of the Soil,* ed. F. E. Bear. New York: Reinhold Publishing Corporation, 1955.

Thorne, W. "The Grazing-Irrigated Region." *The 1957 Yearbook of Agriculture: Soil.* Washington, D.C.: U.S. Government Printing Office, 1957. Pp. 481–94.

Thornthwaite, C. W. "An Approach Toward a Rational Classification of Climate." *Geographical Review* 38, no. 1 (1948): 55–94.

Thorp, James. "How Soil Develops Under Grass." *The 1948 Yearbook of Agriculture: Grass.* Washington, D.C.: U.S. Government Printing Office, 1948.

——. "Effects of Certain Animals That Live in Soils." *Selected Papers in Soil Formation and Classification.* Madison, Wis.: Soil Science Society of America, 1967.

Wadia, D. N. "The Pleistocene System." *The Geology of India.* London: Macmillan and Company, Ltd., 1953. Pp. 398–402.

Walker, N., ed. *Soil Microbiology.* New York: John Wiley and Sons, 1975.

Wasser, C. H., Lincoln Ellison, and R. E. Wagner. "Soil Management on Ranges." *The 1957 Yearbook of Agriculture: Soil.* Washington, D.C.: U.S. Government Printing Office, 1957. Pp. 633–41.

Index

Acid, fulvic, 134, 206
Adhesion water, 45–47
Aeration pores, 44
Aerobic, 27, 192
Albic horizon, 129, 133, 135, 192
Albolls, 124, 125, 192
Alfisols, 74, 149–52, 193
  land use, 155–58
  suborders:
    Aqualfs, 151, 152, 194
    Boralfs, 151, 152, 198
    Udalfs, 151, 152, 224–25
    Ustalfs, 151, 152, 225
    Xeralfs, 151, 152, 226
  U.S. distribution, 144
  world distribution, 144
Amendment, soil, 140, 193
Anaerobic, 8, 27, 193–94
Andepts, 93, 95, 194
Aqualfs, 151, 152, 194
Aquents, 85, 86, 194, 195
Aquepts, 93, 95, 195
Aquods, 136, 137, 195
Aquolls, 124, 125, 195
Aquox, 167, 168, 195
Aquults, 153, 154, 195
Arents, 86, 87, 195–96
Argids, 110, 111, 196
Argillic horizon, 100, 109, 148–49, 196
Aridisols, 74, 99–114, 196

  land use, 111–14
  suborders:
    Argids, 110, 111, 196
    orthids, 110, 111, 212
  U.S. distribution, 101
  world distribution, 101
Atmosphere, soil, 52, 219
Available water, 47–52, 197

Bases (exchangeable bases), 10
Base saturation, percentage, 20, 197
Boralfs, 151, 152, 198
Borolls, 124, 125, 199
Bulk density, 41–42, 199

Calcic horizon, 100, 106–7, 200
Calcification, 106–7
Caliche, 106–7, 200
Cambic horizon, 100, 200
Capillary pores, 44
Capillary tension (potential), 44
Capillary water, 46
Carbonation, 9
Carbon-nitrogen ratio, 39, 200
Carrying capacity, 112
Catena, 55, 200
Cation, 20
  adsorption, 20
  exchange, 20, 200
  exchange capacity, 20, 200

Classification, soil, 67–81
    Marbut System, 68–72
    U.S. Comprehensive System, 72–81
        Family, 73, 78
        Great Group, 73,78, 80–81
        Order, 73–76
        separation criteria, 72–81
        Series, 73
        Subgroup, 73, 78
        Suborder, 73, 77
Clay minerals, 13–14, 179–82, 200, 201
Cohesion water, 46–47
Colloid (*see also* Micelle), 19, 202
Cutans (clay skins), 109–110, 201

Decomposition, 5, 203
Deep tillage, 98
Desert pavement, 110, 203
Desert varnish, 110
Diagnostic horizon (defined), 72
Disintegration, 5
Distribution, of world soils, 75
Dry farming, 128
Duripan, 100, 203

Element (defined), 4
Eluviation, 60, 203–4
Entisols, 73, 83–88, 204
    land use, 96–98
    suborders:
        Aquents, 85, 86, 195
        Arents, 85, 87, 195–96
        Fluvents, 86, 87, 205
        Orthents, 86–88, 212
        Psamments, 86, 88, 216
    U.S. distribution, 85
    world distribution, 84
Epipedon:
    anthropic, 143, 194
    histic, 143, 207
    mollic, 115, 143, 211
    ochric, 100, 143, 212
    plaggen, 93, 143
    umbric, 143, 225

Evaporation, 48
Evapotranspiration, potential, 48–52,
    102–5, 204
Evolution, soil, 59

Fauna, soil, 32–35
    arthropods, 34
    protozoa, 25, 32
    worms, 25, 32–34
Ferrods, 136, 137, 204
Fertility, 11, 204
Fibrists, 174, 175, 204
Field capacity, 47, 204
Flora, soil, 26–31
    actinomycetes, 29–30
    aerobic vs. anaerobic, 27
    algae, 31
    bacteria, 25, 27–28
    fungi, 30–31
Fluvents, 86, 87, 205
Folists, 174–76, 205
Fragipan, 129, 205
Free water, 47
Frost shattering, 7
Fulvic acid, 134, 206

Gilgai, 90, 100, 206
Gleization, 165–66
Gravitational water, 47, 206
Gypsic horizon, 100, 206

Hemists, 175, 176–77, 206
Histosols, 74, 173–78
    land use, 177–78
    suborders:
        Fibrists, 174, 175, 204
        Folists, 174–76, 205
        Hemists, 175, 176–77, 206
        Saprists, 175, 177, 217
    U.S. distribution, 174
Horizon, soil:
    defined, 60, 219
    described, 60–63
Humification, 30, 38, 207

Humods, 137, 138, 207
Humox, 167, 168, 207
Humults, 153, 154, 207
Humus, 38–39, 207
Hydration, 8
Hydrolysis, 8

Illuviation, 60, 208
Inceptisols, 73, 91–96, 208
  land use, 96–98
  suborders:
    Andepts, 93, 95, 194
    Aquepts, 93, 95, 195
    Ochrepts, 93, 95, 212
    Plaggepts, 93, 95, 215
    Tropepts, 95, 96, 224
    Umbrepts, 95, 96, 225
  U.S. distribution, 94
  world distribution, 94
Individual, soil, 59
Ion, 4, 208

Laterization, 164–65
Leaching, 21, 209
Lithic contact, 100, 209
Litter, 35
Loam, soil, 16, 17, 209

Macropore, 44
Magma, 4, 209
Marling, 140, 210
Matric potential, 46
Melanization, 120
Micelle (*see also* Colloid), 19
Micropore, 44
Mineral, 3, 10
  weathering products, 12–17
  weathering resistance, 9–10
Moisture, soil, 44–52, 220
  adhesion, 45–47
  available, 47–52, 197
  capillary, 46
  cohesion, 46–47
  field capacity, 47

gravitational, 47, 206
infiltration, 44
infiltration capacity, 44
regions of the United States, 51
seasonal variation, 51
wilting point, 47, 226
Mollic epipedon, 115, 211
Mollisols, 74, 115–28, 211
  land use, 126–28
  suborders:
    Albolls, 124, 125, 193
    Aquolls, 124, 125, 195
    Borolls, 124, 125, 199
    Rendolls, 124, 125, 216–17
    Udolls, 124, 125, 225
    Ustolls, 124, 125, 226
    Xerolls, 124, 125, 227
  U.S. distribution, 117
  world distribution, 116
Mottles, 9, 211

Natric horizon, 100, 109, 212
Nematode, 32
Nitrogen fixation, 28–29, 212
N-value, 143

Ochrepts, 93, 95, 212
Organic matter, 35–39, 212
  influence on color, 39
Organisms, soil (*see also* Soil, fauna, and Soil, flora), 25–31
Orthents, 86–88, 212
Orthids, 110, 111, 212
Orthods, 137, 138, 212
Orthox, 167, 168, 213
Oxic horizon, 159, 213
Oxidation, 8
Oxisols, 74, 159–72, 213
  land use, 167–72
  suborders:
    Aquox, 167, 168, 195
    Humox, 167, 168, 207
    Orthox, 167, 168, 213
    Torrox, 167, 168, 224

Ustox, 167, 168, 226
world distribution, 160

Pan, 129, 213
Paralithic contact, 100, 213
Parent material, 3–6, 36–38, 213
Particle density, 42–43, 214
Ped, 18, 214
Pedologist, 3, 19
Pedogenic, 50, 214
Pedon:
    defined, 63, 214
    illustrated, 64
Peptization, 134
Percentage base saturation, 20
Percent pore space, 43
Percent soil solids, 43
Petrocalcic horizon, 100, 215
pH, 22–23, 215
Placcic horizon, 129, 215
Plaggept, 93, 95, 215
Plinthite, 153, 155, 159, 165–66, 215
Podzolization, 133
Polypedon, 63–64
Pore space, 41–44, 215
    aeration pores vs. capillary pores, 44
    macropores, vs. micropores, 44
    percentage, 43
Pressure pan, 97–98, 213
Profile, soil, 64–65, 216
    defined, 64–65
    descriptive symbols, 61–62, 183–84
    illustrated, 60, 61
Psamments, 86, 88, 216
Puddled, 14

Range management, 111–14
Rendolls, 124, 125, 216–17
Rhizobium, 28, 217
Rhizosphere, 48, 217

Salic horizon, 100, 108, 217
Salinization, 107–11
Saprists, 175, 177, 217

Separates, soil, 12–14, 220
Sequum, 62
Sesquioxide, 134, 159
Slickensides, 90, 100, 218
Soil:
    amendment, 140
    association, 110, 219
    atmosphere, 52, 219
    bulk density, 41–42, 199
    catena, 55, 200
    classification, 67–81
    color, 23, 185–89
    defined, 2, 218–19
    evolution, 59
    fauna, 32–35
    fertility, 11
    flora, 26–31
    horizons, 60–63, 219
    individual (defined), 59
    loam, 16, 17, 209
    moisture, 44–52, 220
    organic matter, 35–39, 220
    organisms, 25–26
    pan, 129
    parent material, 6, 213
    pedon, 63, 214
    pH, 22–23, 215
    pore space, 41–44, 220
    profile, 64–65, 183, 184
    separates, 12–14, 220
    sequum, 62
    series, 69, 220
    solids, percentage, 43
    structure, 18–19, 220
    texture, 14–17, 221–23
    world distribution, 75
Solum, 63, 223
Spodic horizon, 129, 223
Spodosols, 74, 129–42, 223
    land use, 139–42
    suborders:
        Aquods, 136, 137, 195
        Ferrods, 136, 137, 204
        Humods, 137, 138, 207

Orthods, 137, 138, 212
 U.S. distribution, 131
 world distribution, 130
Structure, soil, 18–19, 220
Subsoiling, 98, 223

Taxa, 72–73
Texture, soil, 14–17, 221–23
Tilth, 39, 224
Time, effect of, 59–65
Topography, 38, 53–58
 effect of, 54–55
 local relief, 55–58
Toposequence, 55, 224
Torrerts, 90–91, 92, 224
Torrox, 167, 168, 224
Transpiration, 48–52
Tropepts, 95, 96, 224

Udalfs, 151, 152, 224–25
Uderts, 91, 92, 225
Udolls, 124, 125, 225
Udults, 153, 154, 225
Ultisols, 74, 151–55, 225
 land use, 155–58
 suborders:
  Aquults, 153, 154, 195
  Humults, 153, 154, 207
  Udults, 153, 154, 225
  Ustults, 153, 154, 226
  Xerults, 153, 154, 227
 U.S. distribution, 146
 world distribution, 145

Umbrepts, 95, 96, 225
Unloading, 7
Ustalfs, 151, 152, 225
Usterts, 91, 92, 225
Ustolls, 124, 125, 226
Ustox, 167, 168, 226
Ustults, 153, 154, 226

Vertisols, 73, 89–91, 226
 illustrated, 91
 land use, 96–98
 suborders:
  Torrerts, 90–91, 92, 224
  Uderts, 91, 92, 224
  Usterts, 91, 92, 225
  Xererts, 91, 92, 226–27
 U.S. distribution, 90
 world distribution, 89

Water budget, 49–50
Water, soil (*see* Moisture, soil)
Weathering:
 chemical, 7–9
 decomposition, 5
 disintegration, 5
 general, 5–12
 mechanical, 7
Wilting point, 47, 226

Xeralfs, 151, 152, 226
Xererts, 91, 92, 226–27
Xerolls, 124, 125, 227
Xerults, 153, 154, 227